Unseen Genders

ERUPTIONS
New Thinking across the Disciplines

Erica McWilliam
General Editor

Vol. 12

PETER LANG
New York • Washington, D.C./Baltimore • Bern
Frankfurt am Main • Berlin • Brussels • Vienna • Oxford

Unseen Genders

Beyond the Binaries

Edited by
Felicity Haynes and
Tarquam McKenna

PETER LANG
New York • Washington, D.C./Baltimore • Bern
Frankfurt am Main • Berlin • Brussels • Vienna • Oxford

Library of Congress Cataloging-in-Publication Data
Unseen genders: beyond the binaries / edited by
Felicity Haynes & Tarquam McKenna.
p. cm. — (Eruptions: new thinking across the disciplines; vol. 12)
Includes bibliographical references (p.   ) and index.
1. Sex role. 2. Gender identity. 3. Sex differences. 4. Human genetics.
5. Sex determination, Genetic. 6. Transsexuals—Identity. 7. Gays—Identity.
I. Haynes, Felicity. II. McKenna, Tarquam. III. Eruptions; vol. 12.
HQ1075 .U65   305.3—dc21   00-034803
ISBN 0-8204-5024-3
ISSN 1091-8590

Die Deutsche Bibliothek-CIP-Einheitsaufnahme
Unseen genders: beyond the binaries / ed. by:
Felicity Haynes; Tarquam McKenna.
–New York; Washington, D.C./Baltimore; Bern;
Frankfurt am Main; Berlin; Brussels; Vienna; Oxford: Lang.
(Eruptions; Vol. 12)
ISBN 0-8204-5024-3

Cover photo, "Chris 1971," reprinted by permission
of the photographer © Chris Somers xxy with the assistance of Geoff Somers
Cover design by Joni Holst

© 2001 Peter Lang Publishing, Inc., New York

All rights reserved.
Reprint or reproduction, even partially, in all forms such as microfilm,
xerography, microfiche, microcard, and offset strictly prohibited.

Table of Contents

List of Figures vii

Introduction—*Felicity Haynes* 1

1. Revealing Hidden Selves 17
Unseen Genders: Looking for the Orlando Effect
  —*Delphine McFarlane* 19
Intersex: Beyond the Hidden A-Genders
  —*Chris Somers and Felicity Haynes* 29
A Sometime Woman: Gender Choice and Cross-Socialization
  —*Michael A. "Miqqi Alicia" Gilbert* 41
Making a Transgenderist: The Construction of Gender Identity
  in Boston and Amsterdam—*Adrianne Dana-Tabet* 51
The Art and Nature of Gender—*Jamison Green* 59
Enactments of Difference—*Tarquam McKenna* 71
Gender Performativity and Normalizing Practices
  —*Wayne Martino and Maria Pallotta-Chiarolli* 87

2. Ways of Understanding Others 121
Tales of the Unexpected: Exploring Transgender Diversity
  through Personal Narrative—*Richard Ekins and
  Dave King* 123

Studying Transsexual Identity—*Katherine Johnson* 143
Gender Love and Gender Freedom—*Surya Monro* 157

3. Toward Theories of Androgyny 167
Gender as (Native) Language—Sam Dylan More 169
Beyond the Binary: Fuzzy Gender and the Radical Center
 —Ashley Tauchert 181
Fractured Masks: Voices from the Shards of Language
 —Lee Anderson Brown 193

Conclusion—Tarquam McKenna 203

Appendix: Internet Links and Resources 213

Glossary 215

List of Contributors 223

Index 227

Figures

Sex/gender attributes 13
Transgendering 124
Binary continua 185

Introduction

Felicity Haynes

> There's a real simple way to look at gender: Once upon a time, someone drew a line in the sands of a culture and proclaimed with great importance, "On this side you are a man; on the other side you are a woman."
> —Bornstein, *Gender Outlaw*, p. 21

The line is literal in many schools where students line up outside the classroom door, girls on one side, boys on another. It becomes a propositional truth when people are required to make a statement by ticking a box, male or female, when applying for jobs, or university admission, making medical, tax, and insurance claims, even buying an airline ticket. But as Kate Bornstein remarks, someone drew that line, and, as she and Anne Fausto-Sterling (2000) demonstrate, it is a mythical one, even in medical science. The binary system requires the Other, on a deficit model in terms of what it excludes or what needs treatment, curing or improving. An essentialist and androcentric binary system establishes a politics of backlash, where boys will be boys (Willis and Kenway, 1997) or, worse, homophobia is justified as a natural instinct to protect reproduction of the species (Bem, 1993).

At the Third International Congress on Sex and Gender held in 1998 at Oxford University, many people demonstrated resistance to the male/female binary. The Bursar of Exeter College, for instance, Susan Marshall, presented as an elegant and confident woman, but she was born a male and has undergone surgery, hormonal treatment, and a legal name change to make her female. At the congress there were intersex people, born with ambiguous genitalia or anomalous chromosomes and morphology (defined medically as "suffering from" such pathologies as adrenal hyperplasia (CAH), androgen insensitivity syndrome (AIS), hypospadias, or gonadal dysgenesis). There was a person with a penis and a uterus, who had breasts until they were surgically removed and now has a beard and male baldness after 27 years' administration of testosterone. There were many men who were indistinguishable in appearance and behavior from each other except that some, having been born female, were unable to marry their female partners. There were males dressed as glamorous women. There were people whose performativities were neither male nor female, or were, if you like, both male and female—gay people, lesbians, bisexuals, and transvestites. The categories of transsexual, intersex, or homosexual that labeled them were not exclusive, but fluid and complex. Some papers in this book arose from that conference.

Nearly all of these people saw themselves as queer, disenfranchised, or pathologized in some way by the prevailing male/female binary. They literally incorporated Judith Butler's (1993) point that what we think of as "sex" is in fact embodied gender; the corporeal incarnation of a discursively constituted (performative) gender. For Butler, the material uninscribed body does not determine gender but the reverse. Gender inscribes bodies. Transsexual people must pass a gender identity disorder psychiatric test before being allowed to have reassignment surgery. They must prove to a psychiatrist that their gendered identity is at odds with their physical morphology, a game which presumes male-female and sex-gender binaries. In many cases they end up "playing the game" of having a psychiatric disorder. Similarly, gays and lesbians who are advised to seek psychiatric counseling for psychological or social deviance do not see a sexual preference for someone of their own gender as unnatural, even if it is less common than heterosexual preference. Most believe that it is time that the silence, even hostility, that surrounds those who do not see themselves as "straight" male or female should be removed.

Anne Fausto-Sterling (2000) uses real-life cases and a probing analysis of centuries of scientific research to demonstrate how scientists have historically politicized the body. She breaks down three key dualisms—sex/gender, nature/nurture, and real/constructed—and asserts that individuals born as mixtures of male and female should not be forced to compromise their differences to fit a flawed societal definition of normality. That is one of the contentions of this book and of Alice Dreger's (1999) latest writing about intersex people. Yet the dualisms persist, inextricably supporting each other.

In the early 1960s, Johns Hopkins University achieved a reputation for surgically reassigning the genitalia of any child born with a clitoris longer than 0.9 cm or a penis shorter than 2.5 cm. On the grounds that it was easier to "make holes than build poles" (Dr. John Gearheart, quoted in Alvarado, 1994) the "micropenis" or large clitoris[1] was usually removed, sometimes with dire consequences for later sexual sensation. Like Judith Butler (1993), one of their leading surgeons, John Money, believed in the constitutive power of gender as a discourse, and the materialization of gender as anatomical sex. He popularized these assumptions with his account of his "success" in raising as a girl an identical twin, David Reimer, who lost his penis in a circumcision accident at age 7 months. Money had recommended that plastic surgery be used to make the boy's genitals appear female, female hormones administered during adolescence to complete the metamorphosis and the boy be told nothing of his birth as a male. Despite Michael A "Miqqi Alicia's" Gilbert's contention (see 41-50) that sex is mainly socially constructed, David, believing himself to be a natural girl, had a very difficult time making friends. His clothes and demeanor, to his peers, did not jibe. Because of his behaviors they called him "caveman" and "gorilla" and would not play with him. None of David's peers knew anything of his genitals. David's story would be familiar to all transsexuals who feel they have been born into the wrong body. No amount of learning the social rules makes them feel as though they fit others' expectations. David (known as John/Joan) recalls: "They kept making me feel as if I was a freak."

Starting at age 14, against the recommendations of the clinicians and family, and without yet knowing of the original XY status, Joan, as much as possible, refused to live as a girl. She preferred jeans and shirt, due to their gender neutral status; and boys' games and pursuits. Joan's daytime fantasies and night dreams during elementary school involved seeing herself "as this big guy, lots of muscles and a slick car and have [sic] all kinds of friends." She aspired to be a mechanic. She refused to look at pictures of nude females she was supposed to emulate. Rorschach and Thematic Apperception Tests at the time elicited responses more typical of a boy than a girl. Her adamant rejection of female living and her improved demeanor and disposition when acting as a boy convinced local therapists of the value of sex re-reassignment (Colapinto, 1997).

It was not just that David was born male and surgically reassigned as female that made his belief in his maleness so strong. Social expectations and surgery were not sufficient to make him a girl. Diamond concluded (Diamond and Sigmundson, 1997) that normal humans are not psychosexually neutral at birth but are biologically predisposed and biased to interact with the environment, that is, familial and social forces, in a male or female mode. Both Money, believing gender to be culturally imposed, and Diamond, believing it to be biologically innate, agreed that the distribution of sex and gender was bimodal, consisting of two mutually exclusive categories, and that anything else was pathological.

If doctors could succeed in transforming hermaphrodites and children with ambiguous genitalia into males and females, then it was equally possible to surgically reassign sex to those who felt as if they had been born into the wrong body, those diagnosed as suffering from "gender dysphoria." The international phenomenon of transsexuals, those who have undergone surgery and hormone treatment to change their sex, is still relatively new but paradoxically it has increasingly challenged assumptions that one's "true" sex is determined by one's genitalia or reproductive organs. While surgeons hold that "correction" of intersex persons can be achieved by making their genitalia more male or female, they can equally feminize or masculinize transsexuals' sex or gender by altering their genitalia to conform to their perceived identity. A new criterion for being male or female is created, one's gender identity, the sex one believes one's self to be, regardless of physiology.

The terms "gender identity" and "gender dysphoria" came to us through medical clinics. It is often hard to realize that the term "gender" has been used only since 1955. Transsexualism shifts the male/female binary to gender as well as sex (King, 1993), failing to challenge the embedded assumption held by most professionals that one must be either male or female. Lesbians, gays, transvestites, male-to-female (MtFs) and female-to-male (FtMs) transsexuals and intersex persons remain medically classified not only as queer, but as still medically pathological, deformed in some way, queer.

This binary model depends on the view that nature can be deemed to make mistakes in gender as well as sex that culture, and especially doctors, can correct. Dreger (1998), using Foucault, described the historical increase in the power of the medical institution to define and determine current standards of gender and sex. Hermaphroditism has been pathologized to such a degree that it is invisible and unthinkable to many doctors as well as lay people and teachers. Hermaphrodites, or intersex persons, are those who are born with physiological characteristics of both male and female. Their number is variously given, but Fausto-Sterling (1993) indicates it may be as high as 1–4% of the population or 1 in 2,000 (Dreger, 1999:14). The difficulty of counting arises not only from the multiple variations (over 70) of morphic evidence of their intersexual status, but from the fact that doctors are required to register all children at birth as either male or female, and therefore usually surgically alter them to make them more male or female, sometimes without their parents' knowledge. To help doctors in this normalization process, Benjamin (1997) published guidelines to help doctors determine which gender they should assign to genetically ambiguous children. Hermaphrodites with Attitude, a

North American society for intersexed people (Kessler, 1998; Dreger, 1999), has argued that the determination of doctors to reassign children as "normal" males or females is often misguided.

One standard measure of a person's gender is chromosomal. A person with 46XY chromosomes is normally defined as male, a person with 46XX chromosomes as female though sex phenotype does not always coincide with chromosomes. There are those whose chromosomes do not fit the binary male/female division. All intersex children diagnosed with Klinefelter Syndrome have 47XXY chromosomes, though the converse does not follow. This account was extracted from the website of the Klinefelter Syndrome Association.

> Klinefelter Syndrome (KS) is probably the most common chromosomal variation found in humans. In random surveys, it is found to appear about once in every 500 live born males. Since the largest percentage of these men would have never been diagnosed otherwise, it shows that in many cases affected individuals lead healthy, normal lives with no particular medical or social questions.
>
> KS is...caused by a chromosome variation involving the sex chromosomes. The person with KS is a male who, because of this chromosome variation, has a hormone imbalance. While Dr. Harry Klinefelter accurately described this condition in 1942, it was not until 1956 that other researchers reported that many boys with this description had 47 chromosomes in each cell of their bodies instead of the usual number of 46.
>
> This extra sex (X) chromosome causes the distinctive make-up of these boys. All men have one X chromosome and one Y chromosome, but sometimes a variation will result in a male with an extra X. This is Klinefelter Syndrome and is often written as 47,XXY.
>
> **Common Characteristics**
> The four most common conditions males with KS may have are sterility, breast development, incomplete masculine body build, and social and/or school learning problems ...
>
> The most common characteristic of men with KS is sterility. Adolescents and adults with KS have normal sexual function but they cannot produce sperm for fathering children. Keep in mind that about 10% of all couples in the general population cannot conceive their own families either.
>
> Frequently, adolescent boys with KS may undergo some breast tissue development. In fact, this is not too much different from boys without KS who also tend to have some breast tissue development during puberty. However, while it tends to be temporary and disappear in other boys, it may persist and the breast tissue may continue to increase in size in boys with KS. In some cases, this may necessitate surgical removal of the breast tissue.
>
> Although most boys with KS are tall (the average is 6' 1/2"), they may not be particularly athletic or coordinated. The penis is usually of average

length, although the testes are small.

Studies indicate an increased risk for speech and language problems which contribute to social and/or school learning problems. Boys with KS may have less confidence in their maleness than other boys. They may be more immature, shy and dependent than their brothers and other boys their age. They may be somewhat passive and apathetic; they may lack initiative, be very sensitive, and have a fragile self-esteem...

Treatment
The major effect of the extra X chromosome seems to be the function of the testes...The most common form of treatment involves administering depotestosterone, a synthetic form of testosterone, by injection once a month...Treatment should result in normal progression of physical and sexual development, including pubic hair growth, an increase in the size of the penis and scrotum (but not the testes), beard growth, deepening of the voice, and increase in muscle bulk and strength.

According to many of the members in our group who are on testosterone therapy, other benefits have included: clarity of thought, better retention of details, more energy and a higher degree of endurance, less hand tremors, better self control, greater sexual drive, easier time in school and work settings, better self-esteem.

To many people this reads like an accurate professional description. Note how the discourse assumes that someone with a chromosomal configuration of 47XXY is male. Could this syndrome appear in live *human* births, as opposed to male births? What is it—a penis or a Y chromosome?—that renders the child an abnormal male? Why is the reading even of 48XXXY or 49XXXXY still given as males with extra Xs, when the default sex is female until the androgens begin to effect physiological change? What is the average penis length for a woman? The possibility of female fertility (of generating ova and bearing a child) in XXYs is not contemplated, nor is the occasional presence of ovaries and uterus, sometimes with a concealed or partial vagina, resulting in disbelief when evidence is presented to doctors.

If one could conceive of XXY persons as both male and female, one could redescribe the social symptoms that constitute KS. Social "problems" such as having "less confidence in their maleness than other boys," being "passive and apathetic," or having "a fragile self-esteem" are problems only for inadequate males. Foucault (1973:93) had noted how naming a symptom in one sense constituted the sign. Could both the fragile self-esteem and the frustration-based outbursts possibly be less a symptom of the syndrome than a resistance to being read as male when one doesn't feel that way?

Many XXYs are both male and female, difficult as that may be for many to believe. Some have penises and uteruses, testes and ovaries.

Resistance to the male gaze is *not* always lessened by surface treatment of the body, as Milton Diamond revealed in the case of John/Joan. Chris Somers, a 47XXY person, supports this. Despite the removal of his/her breasts at the age of seventeen, and the administration of doses of testosterone that allowed him/her to grow a beard, Chris has never felt like a male or a female. The "positive" effect of a testosterone cure is positive only in a normalized male context. Even assuming the male-female binary, to see an XXY person as an abnormal male masks the possibility of reading an XXY body as that of an abnormal female. The female elements of 47XXY persons are repressed in favor of the male, and the "cure" or treatment for the disorder is to adjust the secondary sex characteristics as far as possible to normalized maleness. While the treatment cannot change the chromosomes, it can adjust the hormones.

Does such androcentrism matter? Testosterone therapy, which on the above account gives 47XXY people "an easier time in school and work settings," points to what many feminists have already perceived, namely the normalizing of all humans as male, and that is an issue that feminists have already spent much time trying to redress. But normalizing as male or female may be just as damaging as it was to John/Joan. According to the Intersex Society of North America (ISNA), there are more than 2,000 surgeries performed each year in the United States aimed at surgically assigning a sex of male or female to intersex patients—a type of surgery known as IGM, or Intersex Genital Mutilation. U.S. experts estimate that at least five children are cut every day, mostly making AIS and CAH chromosomal males into genital females. In May and August 1999, the Constitutional Court of Colombia issued two decisions that significantly restrict the ability of parents and doctors to resort to the scalpel when children are born with atypical genitals. The court concluded both decisions with this exhortation:

> Intersexed people question our capacity for tolerance and constitute a challenge to the acceptance of difference. Public authorities, the medical community and the citizenry at large have the duty to open up a space for these people who have until now been silenced...We all have to listen to them, and not only to learn how to live with them, but also to learn from them. (Greenberg, 1999a)

The stories presented on the ISNA website, at the 1998 congress and in this book tell of alienation, ostracization, and isolation, of all those who sit between the binaries, especially intersex persons. To what extent are all of our social institutions responsible for perpetuating the male and female stereotypes? Is the dualism so deeply embedded in our discourse that it cannot be removed?

Abnormal genitalia or secondary sex characteristics are usually the first indications to doctors of "abnormal" sex, and then a chromosomal test will provide the answer as to the "true" sex. Since genetic testing was instituted for women in the Olympic Games, a number of women have been disqualified as "not women" after winning. However, none of the disqualified women is a man; all have atypical karyotypes, and one gave birth to a healthy child after having been disqualified as not being a woman. In this context one's genitalia or morphology are irrelevant. Renée Richards was not allowed to compete with women after she had successful male to female sex reassignment. Billie Jean King, an apparently normal female, proved in the 1960s that she could compete equally with a male tennis player. One could ask whether one's chromosomes should be the defining discriminator or whether the differentiation should not be made simply on the grounds of sporting ability (as with the separation out of the Paralympics, a handicap system based simply on ability), so that those with equal sporting prowess can compete on equal grounds, regardless of identification as male or female.

Though the criteria for identification as male or female shift in different contexts, there is generally an assumption that other gender characteristics will be appropriately present, that is, that male characteristics will cluster together, and so will female characteristics. Schools as institutions do not usually look at karyotypes for gender definition but rely on the birth certificate or parent nomination. Teachers often ignore these and use Christian names, physical appearance, or friendship preferences for gender identification. Yet there are both tacit and explicit rules for being a "proper" girl or boy and schools penalize those who do not comply. In Australia in 1997 a boy was suspended from school for refusing to cut his hair to meet the required dress code of "short back and sides." He appealed on the basis of the antidiscrimination legislation that if it was safe, hygienic, and proper for girls to be allowed to wear their hair long if it was clean and neatly tied in a pony tail, then he should be allowed the same rights. He won on the grounds of sex discrimination. This may seem a more superficial issue than one's identity or sexuality, but it illustrates the fluidity of relevant criteria for maleness or femaleness. Hair became irrelevant as a distinguishing feature.

What is it that makes a boy a boy or a girl a girl? For people with Klinefelter Syndrome it has seemed to be hormonal, "cured" by the administration of testosterone to make them "really" male. For pediatricians it can be the size of the genitalia at birth. In school one's gender ascription may have been made on criteria as simple as choosing to play football rather than netball. For transsexuals, gender identity is the dominant trait, and therefore surgery is recommended to "correct" the anomaly.

Most people, including the surgeons who reassign the sex of intersex children, assume that sex is determined by one's external genitalia, and that is one reason for describing them as primary sex characteristics. As Raymond (1994) showed, the surgeon has as much interest as the patients in having transgendered people desire surgical construction of penises and vaginal reconstruction, and the doctor has as much interest in prescribing hormone treatment to make hermaphrodites appear more male or female. Hospitals and psychiatrists have a vested interest in making people "healthy" and in so doing they perpetuate the norms. Stereotypes of male and female imply that certain gender traits *should* hang together in a coherent bundle and any anomalies need correction.

The International Foundation for Androgynous Studies (IFAS) was established in 1998 to redress such injustices and inform more people about those who share mixed traits of both male and female. IFAS sees the danger in normalizing certain physiognomies as stereotypically male or female and trying to make anyone who appears different meet those norms. It is trying to draw attention to exclusive normalizing practices that pathologize the "queer." As such it includes within its scope of androgyny cross-dressers, transsexuals, intersex persons, gays and lesbians—namely anyone who has been excluded or pathologized by virtue of wishing to exist outside the "norm" of male or female.

This book similarly aims to depathologize those who feel androgynous. It wants to encompass the needs of all those who feel uneasy about being forced to enact a male or female stereotype that does not match their embodied realities. The authors of this book attempt to sidestep binary contests in general. Triangulation serves many purposes but we have no desire to construct a third gender. Rather we wish to educate readers to envisage gender as a complex set of interrelationships that amount to more than an agglomeration of components. Ekins' and King's chapter illustrates the complexity of individual choices that people on the borderlines make to constitute identity within a male/female binary.

Diamond quotes Reiner's 1986 case of an adolescent Hmong immigrant who suddenly dropped out of school at 14 years of age. Although having been unequivocally raised as a girl from birth, the immigrant declared "I am not a girl, I am a boy." Indeed, physical examination revealed a 46XY male hermaphrodite (with mixed gonadal dysgenesis, a female-appearing pelvis, and clitoral hypertrophy). All her school-age friends had been boys. She enjoyed rough and tumble play, avoided dolls and girls' activities, and insisted on wearing gender-neutral clothes. Her feelings of being different —being a boy—developed from about the age of 8 and came to a head at 14 years. Treatment involved surgery and endocrine therapy. This individual, after a period of some depression, progressively developed into a gynecophilic sexually active male and Reiner changed his career from surgery to psychosexual development.

The Hmong immigrant's case, like those in the chapter by Somers and Haynes, reaffirms the point that genital appearance cannot and should not determine sex. This is confirmed by some transsexuals who hoped to become "truly" female after surgery and hormone treatment and are often disappointed that these "surface changes" are not more effective in changing their sex. For most of them their gender identity, their strong belief that they are either female or male, is the defining characteristic of their sex, as it was for David. Still, in some European countries, despite a classification of gender dysphoria, gender identity is not enough either, and many people are not allowed to change their gender unless they also sacrifice their reproductive properties either by castration or by full hysterectomy, apparently on the assumption that their reproductive organs are the essential defining characteristic of sex.

The Hmong immigrant's case also raises a matter that is usually rendered visible only in a context of compulsory heterosexuality, that of erotic orientation or sexual preference. What sense does it make to classify the Hmong immigrant as a lesbian? Or a homosexual? Though erotic orientation has not been used to classify people as male or female, to transgress compulsory heterosexuality is to define oneself as "queer," beyond the normal gaze.

Perhaps if the binary divide was rendered more flexible, gays would not be classified as those in male/male relations or lesbians as those in female/female relations. Even if we used "gynecophilic" or "androphilic" to point more neutrally to the direction of sexual preference, we are still reliant on dominant assumptions of male/female binaries. Why should there be a gender/sex distinction required for sexual preference? Probably because of the need for reproductive survival of the species that originally drove the male-female distinction in the first place. To ask the question as to whether erotic orientation is a biological or cultural matter is once again to remain situated in another gender/sex binary that is not easily maintained. We need not only to step aside from the male-female binary but to step beyond the traditional requirements of necessary and sufficient conditions for use of the sex terms such as male and female, gender terms such as feminine and masculine, and any link between the two. Gender is a complex bundle of characteristics and sometimes it is difficult to hold them all in place.

The first section of this book illustrates how complex this bundle is by presenting stories of those who have resisted categorization as either male or female in many diverse ways, and of their isolation and invisibility in a society that pathologizes and excludes them.

The retreat from the gender binary that has dominated cultural forms in the modern period is an ethical project. It is not possible to ignore the binary, but it is possible to engender models for sexual embodiment that re-constitute normativity. Since the 1970s, Sandra Bem (1972, 1993) has been devising tests that indicate a psychological androgyny in most people, that is, that they combine psychological traits traditionally conceived of as male or female.

Recognizing the actuality of a physiological and psychological androgyny may require us to revise our system not only of permissible sexual relations, but of public toilets, and our assumptions of marriage and family. Cranny-Francis (1995) analyzes the political motives underlying all binary terms, establishing that to destroy those binaries is a subversive act. But some subversive activity is necessary if any change is possible. Let us be subversive.

The essays in this book speak to the questions that emerge from a poststructuralist critique of binary logics, like that offered by Butler and Scott (1995): What happens to gender binaries when traditional grounds and foundations—experience, history, universal norms—are called into question? How does such questioning open up possibilities for reformulating agency, power, and sites of political resistance? Can androgyny problematize the notion of the subject without losing its political effectiveness? What are the consequences of an androgynous reformulation of difference? What implications could a poststructuralist critique of the binary logic of male and female and of sex and gender have for the theorization of racial and class differences as well as gender differences?

If we accept the existing centralities of binary oppositions in any language or culture we can, as Tauchert (see 182) argues, change them in three ways. First, we can invert the weighting and values of the binary. This means replacing the androcentrism in the definition of Klinefelter Syndrome with gynocentrism. This causes backlash and resistance. Another strategy might be the move adopted in queer theory: to displace the two existing binaries of male and female by creating another "third gender"—the queer or androgyne. This in turn sets up its own binary—the it's, versus the he's and she's—and ends up perpetuating a new stereotype which in turn provokes opposition and backlash. The third strategy, adopted in this book, is to undermine the binary process as a language game imposed on a more general de-hierarchized "web-sphere of gender differences."

Ashley Tauchert's chapter challenges the assumption that we need to polarize categories in order to separate and clarify our positionality, and ultimately our identities. Tauchert places the androgyne at the radical center. Her "fuzzy" model takes the inter-sexed subject as the normative human subject, to which all embodied subjects conform to some degree. By recognizing sexual complexity, it is possible to enfranchise transsexuals, bisexuals, transvestites, gays, lesbians, and the invisible but numerous intersex persons or hermaphrodites. In other words, we wish to provide a model that does not require one to be either simply male or simply female.

There have been cultures in which the line between male and female is not so firmly drawn: for example, the *hjira*s of India, and the *berdache* of Native America. Alice Dreger's (1998) history of hermaphroditism contrasts the twentieth-century marginalization of the hermaphrodite and the homosexual with the more inclusive gender assumptions of the pre-Socratics. Her analysis reveals that the criteria for sex or gender shift, often so subtly that we assume the criteria are natural and essential. It is time to place them in context and simply see how and for what purposes "male" or "female" is being used.

We shift our criteria for certain purposes, both in the language games we play and in our daily practices. Wittgenstein uses a metaphor of a thread to show us why this is so:

> Consider, for example the proceeding that we call "games." I mean board-games, card-games, ball-games, Olympic games, and so on. What is common to them all?—Don't say: "There must be something common or they would not be called 'games'"—but look and see whether there is anything common at all, but similarities, relationships and a whole series of them at that. To repeat: don't think but look!—Look for example at board games, with their multifarious relationships. Now pass to card-games; here you may find many correspondences with the first group, but many common features drop out, and others reappear. When we pass next to ball-games, much that is common is retained, but much is lost—Are they all "amusing"? Compare chess with noughts and crosses. Or is there always winning and losing, or competition between players? Think of patience...And we can go through the many, many other groups of games in the same way; can see how similarities crop up and disappear.
>
> We see a complicated network of similarities overlapping and criss-

crossing...And we extend our concept(s)...as in spinning a thread we twist fibre on fibre. And the strength of the thread does not reside in the fact that some one fibre runs through its whole length, but in the overlapping of many fibres. (Wittgenstein, 1953: paras 66–7).

Five strands underlying sexual identity/behavior have been identified—genetic (chromosomal), physical appearance, functional structure of the hypothalamus, sexual orientation, and gender identity. One's sexual identity is usually believed to remain constant from childhood but an individual could function as male or female to varying degrees independently in each of these five categories. I add a few more threads below to indicate how gender can be viewed as multidimensional and nonlinear, reminding the reader that these are not exhaustive. The essays, stories, and references in this book show how complex and fluid are matters of sex and gender.

	Male	Female
external genitalia	penis	vagina
reproductive organs	testes	ovaries
physical shape	slim hips, broad shoulders	broad hips, slim shoulders
chromosomes	46XY	46XX
voice, hair, fat	beard	high voice
hormones	testosterone	estrogen
ethical stance	rights	responsibilities
sexual preference	females	males
dress preference	trousers	skirts
performativities	assertive	passive
gender identity	masculine	feminine

Figure 1. Sex/gender attributes

Gender identity is in most cases uncontested, because there is usually a cluster of dominant features that tend to one side or the other, but the normal person will not be entirely male or female. Males and females are here at opposite ends of a continuum without a sharp dividing line. Because sexes and genders are incorrigibly plural, even the attributes do not slide along a unidimensional line. It might even be preferable to see each of the continua listed above as a flat disk, meeting at the androgynous center and radiating out to a sphere with male-female edges forming a tennis ball surface image, a three-dimensional fuzzy Yin-Yang. As Tauchert describes (see 186) as night and day gradually merge through dawn and dusk, so in each gender feature disc, there is a male/female axis at right angles to the defining androgynous fuzzy area so that the features, scattered at various points in the inside of the sphere, will usually be mixed androgynous. Any one person will be a complex mixture of various interdependent male or female components, some culturally attributed or chosen and some naturally acquired.

Now when people are asked to tick a male or female box, they can respond—why do you want to know? Which of these criteria is relevant to your purpose?—and select accordingly, cutting a slice or cross-section through the appropriate features of the gendersphere. Gender and sex thus become contextualized, and gender identity fluid. If we need to identify a person as a male or female for security purposes, for instance, when required to apply as male or female for an airline ticket, external morphology and dress are the salient features. For childbearing, the reproductive organs are crucial. For establishing a family unit, even where the raising of children is a consideration, genitalia and reproductive organs seem less critical than mutual responsibility and commitment, though this rubs against many popular beliefs. One's reading and spatial abilities may be significantly different depending on one's sex, but for measurement of learning, one's genitalia are irrelevant. There is a world of difference between proposing that the central problem of meaning is "what do we call it?" and proposing that the problem is "how are we supposed to treat it?" We label people as male or female for certain purposes, and the differences in our reasons for their categorization make different features of gender difference salient. Differences both reveal and conceal relationships of value.

One project of this book was to reassert the "natural" in our determination of categories, even in a postmodern epistemology, to reinstate the identity of hermaphrodites as natural persons, not freaks. The world is crazier than we think and incorrigibly plural. In looking at how hermaphrodites came to be instantiated in the world as the Other, we found that many others with overlapping strands of similarity and difference revealed themselves through their stories. There is not a simple third gender of androgyne, but as we shall see a very complex network of similarities and differences.

The first section of this book "Revealing Hidden Selves" presents the stories of many people who have been made invisible in the social arena by the male/female gender binary. The second section, "Ways of Understanding Others," is more like an exposition of research methods, exploring different ways of making meaning from the texts of those who speak as individuals. Neither gender nor sex is free from some natural origins. Gender and sex inform and are informed by our social practices and our understanding of what is real. The third section, "Toward Theories of Androgyny," therefore, raises deeper questions about the relation between our experiences and language and our constitution of subjects, the ways in which we substantiate our experiences of the world and people in it.

There are natural physical variations in the normal male or female body that should be accepted, and there are many threads that constitute not only gender but humanness. As humans we have an ethical mandate to accommodate anyone who feels uncomfortable being strait-jacketed into gender stereotypes by all or any of these features. That includes transgendered people, homosexuals, bisexuals, anyone who feels uncomfortable with the sharp distinction drawn between male and female on any one of the continua of gender attributes. This book presents some of the complexities of sex and gender, deconstructing the "normal assumptions" of several deeply embedded binary oppositions—including body and mind, science and culture, sex and gender, queer and normal, and male and female.

At least three consequences may follow from raising awareness of unseen genders. People may become more accepting of themselves as they are. They may be more easily able to tolerate differences in others, to understand, accept and even celebrate them. People may feel freer to choose how they present themselves, not only as hermaphrodites or transgendered persons able to walk openly down the street with beards and breasts without fear of ridicule, but through surgery to change those features that make them unhappy. Then again, they may not feel as unhappy with features that earlier did not fit social stereotypes. This prefigures a plasticity of identity consistent with both a postmodern world and a predetermined identity. The making of difference between male and female is a serious business. When we do it we should at least be clear why we are doing it and what the consequences are of our categorization of those who may not easily fit our categories. This book should help us reconceptualize our genders.

References

Alvarado, Donna. (1994). Intersex. *San Jose Mercury News*, July 10.
Bem, Sandra. (1972). The measurement of psychological androgyny. *Journal of Consulting and Clinical Psychology*, 42:155-162.
Bem, Sandra L. (1993). *The lenses of gender: Transforming the debate on sexual inequality*. New Haven, CT: Yale University Press.
Benjamin, H. (1997). Sex reassignment at birth: Longterm review and clinical implications. *Archives of Pediatrics and Adolescent Medicine*, 151:1062-1063.
Bornstein, Kate. (1994). *Gender outlaw: On men, women and the rest of us*. New York: Routledge.

Butler, Judith. (1993). *Bodies that matter: On the discursive limits of sex.* New York: Routledge.
Butler, Judith and Joan W. Scott. (1995). *Feminists theorize the political.* London and New York: Routledge.
Colapinto, John. (1997). The true story of John/Joan. *The Rolling Stone*, December 11, 1997. 54–97.
Colapinto, John. (2000). *As nature made him: The boy who was raised as a girl.* London: Harper Collins.
Cranny-Francis, Annie. (1995). *The body in the text.* Carlton, Victoria: Melbourne University Press.
Diamond, Milton, and H. Keith Sigmundson. (March, 1997). Sex reassignment at birth: A long term review and clinical implications. *Archives of Pediatrics and Adolescent Medicine,* 151:1064.
Dreger, Alice. (1998). *Hermaphrodites and the medical invention of sex.* Cambridge, MA: Harvard University Press.
Dreger, Alice Domurat (ed.). (1999). *Intersex in the age of ethics.* Hagerstown, MD: University Publishing Group.
Fausto-Sterling, A. (1993). The five sexes: Why male and female are not enough. *The Sciences:* 20–25.
Fausto-Sterling, A. (2000). *Sexing the body: Gender politics and the construction of sexuality.* New York: Basic Books.
Foucault, Michel. (1973). *The birth of the clinic.* London: Tavistock Publications. Reprinted 1994 by New York: Vintage Books.
Greenberg, Julie A. (1999a). Colombia high court limits surgery on intersexed infants. http://www.isna.org/Colombia/.
Greenberg, Julie A. (1999b). Defining male and female: Intersexuality and the collision between law and biology. *Arizona Law Review* 41(2):265–328.
Herdt, Gilbert (ed.). (1994). *Third sex, third gender: Beyond sexual dimorphism in culture and history.* New York: Zone Books.
Kessler, Suzanne. (1998). *Lessons from the intersexed.* New Brunswick, NJ: Rutgers University Press.
King, Dave. (1993). *The transvestite and the transsexual: Public categories and private identities.* Brookfield, VT: Ashgate Pub. Co.
Laqueur, Thomas. (1990). *Making sex: Body and gender from the Greeks to Freud.* Cambridge, MA: Harvard University Press.
Raymond, Janice. (1994). *The transsexual empire: The making of the she-male.* New York and London: Teachers' College Press.
Willis, S., and J. Kenway. (1997). *Answering back.* St. Leonards, NSW: Allen and Unwin.
Wittgenstein, Ludvig. (1953). *Philosophical investigations.* Oxford: Blackwell.

Notes

1   If that question seems absurd, there are many examples from Renaissance literature that reveal the clitoris to be considered a female penis (Dreger, 1998; Laqueur, 1990).

1. REVEALING HIDDEN SELVES

> Our most stubborn and pertinaceous assumptions are precisely those which remain unconscious and therefore uncritical...concepts we take for granted without realizing that we do so at our peril...the best and perhaps the only sure way of bringing to light and revivifying our fossilized assumptions, and of destroying their power to cramp and confine us, is by subjecting ourselves to the shock of contact with a very alien tradition.
> Harold Osborne—*Aesthetics and Art Theory*

In this section, we hear of different experiences of those who do not easily fit the gender binaries—transsexuals, cross-dressers, gay and lesbian people, intersex people. Chris Somers and Felicity Haynes present narratives of those people intersex who have had to conceal their natural differences in the face of hostility and disbelief. The other chapters present a view of gender not so much as a natural given but as a language which is not easily heard, a performance which is not often exposed to view. Tarquam McKenna speaks of the acting out of identity, of what performances are allowable and what not, especially among gay teachers.

The stories illustrate how the covert silencing of any person (mentally or physically) allows society to imagine that it has eliminated the individual's difference; but, as Delphine McFarlane argues, the silenced individual is made further at risk of estrangement and oppression. This form of gender oppression through hegemonic invisibility is painful for those rendered invisible. There is an unknown number of genders. Those that are culturally visible, male and female, are given identities as they are known to "exist." Those that are not visible in terms of acceptable cultural constructs do not "exist." The "unseen" genders are those that are unimagined and unknown by most people, as is sexuality. Invisible genders are by definition alienated. Alienation is suppression and oppression, forced compliance and containment of the individual.

Jamison Green believes gender preexists social manipulation. Certainly gender *is* used as a tool to oppress people, to control who has power and authority, but its power symbols and their meanings vary across time and between cultures. Green warns against the predominantly feminist critique of gender, saying that it is positioned firmly on a platform of opposition to a particular form of male dominance, patriarchy, and without this grounding principle from which feminism flings its barbs, its theories bear little relevance to individual experience of gender. Foreshadowing Sam More, he asserts that gender is a type of language to which everyone has varying degrees of access. Our access to gender is equivalent to the power to speak.

Michael A. "Miqqi Alicia" Gilbert speaks of the cross-dresser who treats gender as something that is put on and taken off, as opposed to the transsexual who permanently adopts (or has always adopted) the "opposite" gender. Gilbert asks to what extent Social Construction Theory can support an activity of this nature, especially if it ignores the possibility of intentionality.

Adrianne Dana-Tabet describes the influence of the political and social context on conceptions of personal identity and group affiliation in cross-dressing members of a Dutch transformation center in Amsterdam and a transgender community in Boston. The predominant gender discourse in Holland, where the concept of transgenderism is relatively insignificant, suggests a dyadic perception of gender role and identity. Dutch gender and sexually dysphoric individuals are either oriented toward transsexualism and eventual government-supported sexual reassignment surgery or relegated to the unsubsidized, socially marginalized category of fetishistic cross-dresser. In the United States, transgenderism, as an expandable category, acts as an umbrella concept, which subsumes and accommodates a spectrum of gender, sexual, and social identities and communities.

Through interviews with adolescent boys and written responses by girls and transgendered persons, Wayne Martino and Maria Pallotta-Chiarolli examine the normalizing regimes of practice that impact on the various ways in which boys and girls at Australian schools define desirable forms of masculinity and femininity. Compulsory heterosexuality and gender duality prescribe appropriate behavior for boys and girls.

Unseen Genders: Looking for the Orlando Effect

Delphine McFarlane

Denial is a human action that serves ignorance. It is also a defense against discomfort and pain. A society that denies change and difference is a stagnant and damaging society, yet the security of denial is seductive and comfortable. We want everyone to be like us, but who are we and who do we represent? That misused pronoun can be read as the "we," meaning "I" the author, who would like to represent society in the truly egocentric way that we all tend to adopt. "We" would all feel so much more comfortable if everyone else was just like us—or would we? Perhaps our own sense of self is reassured by difference and as long as we can deny the existence of others our own certainty will remain unchanged and unchallenged. The "Me Tarzan! You Jane!" approach has a simplicity that is a safe, no risk approach, but the damage of such strait-jacket judgment is evident in the maelstrom of social inequity in groups and subgroups denied a place, denied a voice. The ultimate result of such alienation is manifest ignorance. We deny the uniqueness of those we name as "others," those "not like us" at our own peril because in reality they are us; we all belong to the tribe that is humanity. Surely the strength of humanity is its multiplicity. The ambiguity of gender is enunciated by Judith Butler (1990:16) in the following statement:

> Gender is a complexity whose totality is permanently deferred, never fully what it is at any given juncture in time. An open coalition, then, will affirm identities that are alternately instituted and relinquished according to the purposes at hand; it will be an open assemblage that permits of multiple convergences and divergences without obedience to a normative *telos* of definitional closure.

Problems can arise when scientists use their positions of power to affirm the identities of infants too young to know and, it would appear, too young to be assessed as definitively male or female. The outward appearance of genitalia is not necessarily congruent with the innate gender of the child or adult concerned. "People are too varied and complex to be 'known' solely through their (*physical*) gender identity" (Query, 1997:26). Surely it is the "situational knowledge" of the particular human being concerned that counts; individuals alone can best judge the appropriateness of the bodies they inhabit. Elizabeth Sourbut (1996) has challenged the hierarchical construct of science, which seeks to deny the complexities of sex and gender in a quest to control and maintain the appearance-related dichotomy of gender, rather than developing an openness to understand the experiences of individuals. Jennifer Gonzalez has cited the fluidity of existence as portrayed through Virginia Woolf's protagonist, Orlando, who changes gender and who adapts to social and cultural changes across many centuries and continents. Orlando may be seen reflected in the nature of the cyborg in the challenge presented by Haraway (1997) and others to contemporary practices of enforced gender roles. Technological advances in medical science have created the freedom for people to change their physical appearance in both radical and subtle ways, and the interconnectedness of the body and technology is becoming endemic in Western cultures. While for some of us this may mean a pin stabilizing a bone or capped teeth, for many transsexual people this means an opportunity to be the gender one is, physically as well as psychologically. Sandy Stone (1998:298) sees this stage of our development in a strikingly liberating manner:

> Here on the gender borders at the close of the twentieth century, with the faltering of phallocentric hegemony and the bumptious appearance of heteroglossic origin accounts, we find the epistemologies of white male medical practice, the rage of feminist theories, and the chaos of lived gendered experience meeting on the battlefield of the transsexual body: a hotly contested site of cultural inscription, a meaning machine for the production of ideal body type. Representation at its most magical, the transsexual body is perfected memory, inscribed with the "true" story of Adam and Eve as the ontological account of irreducible difference, an essential biography that is part of nature. A story that culture tells itself, the transsexual body is a tactile politics of reproduction constituted through textual violence. The clinic is a technology of inscription.

"Textual violence" is a term worth using; it can be used as a tool, as a weapon, as a source of discovery. It is but one of the glut of insightful phrases put to work in the above passage from Sandy Stone: "the faltering of phallocentric hegemony" tumbles before our ears in its evocative alliteration, as does "white male medical practice"; the "rage of feminist theories" is scalpeled into a neat phrase, to make way for the rich exuberance painted into the wordy chaos of "lived gendered experience meeting on the battlefield of the transsexual body." Stone's incisive use of language creates both imagery and aural atmosphere in the quest to identify the lived experience as the essence, the experiential reality. Similarly Judith Butler (1997) suggests the power of the performative through "excitable speech," which interrupts and redirects oppressive language in a counter-force of aggressive reappropriation.

Like Stone's "textual violence," Sourbut's "situational knowledge," and Butler's "excitable speech," the voice of the people, not just white male doctors, will be heard.

The human body is a statement—it says here I am, this is the physicality of my genetic blueprint for existence—I am a unique collection of cells manifesting themselves in this particular physical and psychological form. You can see me—I am real; you can touch me—I have substance; you can speak to me and I will respond—I hear you; and you can hear me. So I now call upon you to hear my voice as I tell you the story of my body, which is me, and my search for a place where I can exist.

David is an Australian man who was raised as a female, but challenged that gender status during puberty. He said to me, in an oral interview:

> I was born a male and named as such. After a couple of weeks of not feeding properly some tests were done and it was found that I had congenital adrenal hyperplasia. At about $1^1/_2$ years I was taken to Princess Margaret Hospital for "investigative surgery", I don't know what inspired them to look inside—I assume my penis wasn't normal. About seven months later they cut it off.
>
> As far as I can guess, doctors presumed my penis was an abnormality, when it could equally have been construed it was very abnormal for a bloke to have a uterus and no vagina. Anyway, they decided for my parents and me I was going to be raised as a girl.
>
> The difficulty of having to behave like a girl through school wasn't a very enjoyable experience, and I won't dwell on it. Suffice to say, if a normal 2-year-old boy lost his penis in an accident it'd be inconceivable to expect him to put on a dress and behave like a girl for the rest of his life—which is what I feel I was expected to do.
>
> At about 14, I was sent to a girls' boarding school, which in circumstances would be a bloke's dream come true, but having to constantly resist the temptation made it a bit of a nightmare really. One of the most embarrassing moments was when I was reading aloud in class and my voice broke. Having to do swimming classes when it was quite obvious I didn't have a girl's body wasn't a lot less humiliating.
>
> At some stage in the next year or so I was put in hospital where I believe I was to be made more like a female. A doctor I hadn't met before came in to tell me what they planned to do. That was when I said I didn't

want that. He left the room without saying anything, and the endocrinologist came down a bit later and said he was going to refer me to a psychiatrist. That was the ultimate insult. Having first been denied by doctors my physical sexuality, it seemed my mental state was also going to be determined.

During my stay at the Sir Charles Gardiner Hospital's D block (D for demented, I was sure) I was told by a nurse she was taking me for a test somewhere. Upon asking what sort of test she said an EEG and I'd have to ask the doctor (psychiatrist) if I wanted to know what for (doctors were still expecting me to ask them what the hell was going on). Anyway, at the electroencephalograph place, after having all the wires stuck to my head, I was told to relax and look at some damn picture on the wall. I didn't know what any of this was about, so I looked anywhere but at the picture and hoped the whole lot of them would drop dead and leave me alone.

About a year later I had surgery, which I suppose can only be described as a hysterectomy.

The endocrinologist referred me to a plastic surgeon. At [when I was] about 18 he and a urologist constructed from my forearm a replacement for what I lost at two. After the operation the surgeon told me there was a baby in PMH with the same problem I'd been born with—they were going to cut his off too. From my experience I thought that'd be the wrong thing to do.

I expect doctors, like most other people, don't make mistakes intentionally. Cutting off the part of the anatomy which distinguishes a boy from a girl seems a very confident decision to make, especially if there's no outward indication the child's a girl.

Since my initial operation, no doctor involved has contacted me to explain why they thought it was necessary, or find out if it was the right thing to do to me—which it wasn't.

I expect it's all been fairly difficult for my parents, brothers, relatives, and everyone I've known, to come to terms with my past, and [they] probably had, and have still, as many doubts and unanswered questions as I did.

But you don't hear the real me, and you don't see the real me. I present myself in a certain manner—a "fitting" manner, so you will feel comfortable with me. I speak in an appropriate voice, saying acceptable things for the sake of stress-free social intercourse. The real me has been "othered" by you and your heterosexual hegemony. I do not have a place other than one that is in hiding; I have been bound, gagged, and masked.

David was able to break the bondage of the inappropriate gender assigned to him by professionals who saw only parts of a physical body, not a total person. That he was able to resist the medical recommendations is testament to his inner strength and conviction regarding his identity. Using his own brand of "textual violence," David clearly enunciates his frustration with, and recognition of, medical presumption in the statement: "it could equally have been

construed it was very abnormal for a bloke to have a uterus and no vagina." Such clarity questions the authority of a science far removed from the people it purports to be helping and out of touch with the complexities of physical reality.

In her search for the materiality of the body, Butler (1994:ix) found that the very subject of materiality moved her into other domains:

> I could not fix bodies as simple objects of thought. Not only did bodies tend to indicate a world beyond themselves, but this movement beyond their own boundaries, a movement of boundary itself, appeared to be quite central to what bodies "are." I kept losing track of the subject. It proved resistant to discipline. Inevitably, I began to consider that perhaps this resistance to fixing the subject was essential to the matter at hand.

Resistance to fixing the subject is the response to attempts to deny existence to physical and sexual forms and choices not characteristically heterosexual. However, resistance is problematic through a discourse that is based upon assumptions of strict male/female roles and scientific superiority, as David observes when he states: "they decided for my parents and me I was going to be raised as a girl." It is only since the challenges made by feminist and queer theory that such resistance to patriarchal values has found a place. While there have always been contradictions to the perceived cultural norms of female and male, these "hidden differences" and "other ways of being" have not previously had a voice or place within societal conventions. We can hear and understand David's rational logic: "if a normal 2-year-old boy lost his penis in an accident it'd be inconceivable to expect him to put on a dress and behave like a girl for the rest of his life." However, such logic was not evident when David's future was under consideration by the experts at the time.

The voices of those "othered" are still having great difficulty being heard, partly because they are compelled to theorize as a counterculture while assuredly fixed within the discourses of the other culture's regime of power; the very processes that reiterate and regulate social norms. Their theories and identificatory processes may be heard, yet they are still alienated because their delegitimized bodies fail to count as "bodies." However, the uncertainty of cultural norms based upon excluding "others" is that the identification of difference actually acknowledges the existence of those for whom there is no place. So while David was forced to comply with the medical interference with his given body, he was ultimately able to redirect the course of his life through establishing his power as an individual and firmly citing his right to be his natural gender.

What is "forced" by the symbolic, then, is a citation of its law that reiterates and consolidates the ruse of its own force. What would it mean to "cite" the law in order to produce it differently, to "cite" the law in order to reiterate and co-opt its power, to expose the heterosexual matrix and to displace the effect of its necessity?

> The process of that sedimentation or what we might call *materialization* will be a kind of citationality, the acquisition of being through the citing of power, a citing that establishes an originary complicity with power in the formation of the "I." (Butler, 1994:15)

It is suggested that such incitement may serve to expose the interconnectedness of femininity and masculinity; the telling of stories to enable others who would explore gender and sexuality beyond assumed roles which are for many people stultifying and alien to their given nature. As Donna Haraway's (1991) cyborg metaphor suggests that the opposition between the organic and the machine is being broken down to the point at which the human being and technology are inextricably conflated, so too may we acknowledge the blending of female and male that is within all bodies. We must welcome the unseen genders and their uniqueness expressed in varying degrees and a rich array of physical and psychological manifestations. We must challenge the science and medicine that would impose outmoded beliefs based upon religious doctrine and conveniently suited to the dominant male of the species.

Sandy Stone (1998:301) states:

> Under the binary phallocentric founding myth by which Western bodies and subjects are authorised, only one body per gendered subject is "right." All other bodies are wrong.

As clinicians and transsexuals continue to face off across the diagnostic battlefield that this scenario suggests, the transsexuals for whom gender identity is something different from *and perhaps irrelevant* to physical genitalia are occultated by those for whom the power of the medical/psychological establishments, and their ability to act as gate-keepers for cultural norms, is the final authority for what counts as a culturally intelligible body. This is a treacherous area, and were the silenced groups to achieve voice we might well find, as feminist theorists have claimed, that the identities of individual embodied subjects were far less implicated in physical norms, and far more diversely spread across a rich and complex structuration of identity and desire, than it is now possible to express.

In her quest for gynogenesis, Sourbut (1996:227–241) also calls for a thinking beyond the cultural and scientific imposition of existing binary categories of female and male which insist upon rigid, outdated, and stereotypical versions of sexuality and gender. The difficulty and the skill lies in reconstructing boundaries and deconstructing dichotomies.

This approach suggests that we must change the ideological system within which science is carried out and also the processes by which knowledge is accumulated, and indeed the definitions of what counts as knowledge. These things all go together: a science that sees itself as being at the top of a hierarchy, that aims to control the natural world, will have no qualms about using the things it studies. On the other hand, a science that seeks to break down boundaries and sees knowledge as situational and dependent upon that which it studies is likely to be more respectful.

Science and the definitions of what counts as knowledge are locked into our language as surely as we are. All expressions of scientific research and discoveries are communicated through written and oral words and figures. The historicity of this language is both our prison and our means of escape. Our reality is made possible by language; we are named or not named; we may choose to speak or not speak; to speak out or to speak within. By speaking out, I suggest we speak outside the acceptable and beyond the boundaries. I suggest we tell the stories of pain and tragedy, of oppression and fear. I suggest we enable the oppressed to speak their stories so that through the enablement we will hear their truth and their strength. And we will discover that the oppression is defeated by the telling.

Donna Haraway (1997:33) makes two claims regarding contemporary science: (1) There have been practical inheritances, which have undergone many reconfigurations but which remain potent; and (2) the stories of the Scientific Revolution set up a narrative about "objectivity" that continues to get in the way of a more adequate, self-critical technoscience committed to situated knowledges.

The manifestation of scientific superiority that would impose unrealistic gender roles upon infants is neither desirable nor practical. David's story and the clear observations of injustice it relates is not simply powerful prose. The scientific knowledge of a doctor cannot possibly deny the situational knowledge of David's experience. Scientific objectivity is not possible due to inherent language bias and to the dubious notion of "objectivity" itself, because when one asks the questions "What is male? What is female?" there are no straight answers; all definitions are eclipsed by the situational knowledge of the person answering.

Stone incites all people who have endured not only radical surgery but often years of psychological torment to let the world know their stories. Her message (Stone, 1998:302) to all transsexuals at the end of her "Empire Strikes Back" paper, implores:

> I ask all of us to use the strength that brought us through the effort of restructuring identity, and that has also helped us to live in silence and denial, for a re-visioning of our lives. I know you feel that most of the work is behind you and that the price of invisibility is not great. But, although *individual* change is the foundation of all things, it is not the end of all things. Perhaps it is time to begin laying the groundwork for the next transformation.

What do we fear when matters of sexuality and gender are brought out into the open? Is it the guilt born of childhood culture sculpting that beats the boy/girl divide into our infant brains? Is it that insidious Oedipal ogre that haunts our soul or is it the fear that our race will abandon us if we are too different, too extreme? Is gender conformity so attractive or is it a response to the unknown and therefore frightening alternatives? Why do we admire alternative heroes and yet cling to so-called normality when we know that the "norm" is just a line in a bell curve? Why have no doctors explained to David why it was imperative to remove a vital and healthy part of his body when he was aged 2 years, when 16 years later they agreed to redress the original surgery?

Let us hear the real people speak, not the statistics, not the scientists, but the previously silenced: the men, the women, the epicene people, the transsexuals, the polygendered; those who have life stories to tell us. Their stories are the data of experiential truths. These truths are the day-to-day normalities of those born not strictly male, not strictly female, or psychologically one gender feeling misplaced in the body of a different gender; unique blends of femininity and masculinity and wonderful combinations of all that it is to be human. Let their "textual violence" cut through the rhetoric of scientific justification. Let our children be free of imposed gender roles and let us all be free to explore the Orlando effect should we so desire.

References

Bordo, Susan. (1990). Feminism, postmodernism, and gender-scepticism. In *Feminism/postmodernism*, ed. Linda J. Nicholson. New York: Routledge, 133–156.

Boswell, John. (1994). *Same sex unions in premodern Europe*. New York: Random House.

Butler, Judith. (1990). *Gender trouble: Feminism and the subversion of identity*. New York: Routledge.

Butler, Judith. (1994). *Bodies that matter: On the discursive limits of "sex."* New York: Routledge.
Butler, Judith. (1997). *Excitable speech: A politics of the performative.* New York: Routledge.
Connell, R. W. (1987). Sexual ideology: Discourse and practice. In *Gender and power.* Sydney: Allen and Unwin, 241–258.
Gatens, Moira. (1996). *Imaginary bodies: Ethics, power and corporeality.* London: Routledge.
Gray, Chris Hables (ed.). (1995). *The cyborg handbook.* New York and London: Routledge, 1995.
Haraway, Donna. (1991) *Simians, cyborgs, and women: The reinvention of nature* New York: Routledge.
Haraway, Donna.. (1997). *Modest_witness@second_millenium.femalemanmeets _oncoMouse: Feminism and technoscience.* New York and London: Routledge.
Lloyd, Genevieve. (1984). *The man of reason: "Male" & "female" in Western philosophy.* London: Routledge.
Lumby, Catharine. (1997). *Bad Girls: The media, sex and feminism in the 90s.* St. Leonards, NSW: Allen and Unwin.
Oudshoorn, Nelly. (1996). A natural order of things? Reproductive sciences and the politics of Othering. In *Futurenatural nature, science, culture,* ed. George Robertson, Melinda Mash, Lisa Tickner, Jon Bird, Barry Curtis, and Tim Putman. London and New York: Routledge, 122–133.
Query, Julia. (1997). Candid Cameron. *Women's review of books,* vol. 14, 26–27.
Sourbut, Elizabeth. (1996). Gynogenesis: A lesbian appropriation of reproductive technologies. In *Between monsters, goddesses and cyborgs: Feminist confrontations in science, medicine and cyberspace* ed. Nina Lykke and Rosi Braidotti. London: Zed Books 227–241.
Stone, Sandy. (1998). The Empire strikes back: A posttranssexual manifesto. In *The visible woman,* ed. Paula A. Treichler, Lisa Cartwright, and Constance Penley. New York and London: New York University Press 287–310.
Tasker, Yvonne. (1993). *Spectacular bodies: Gender, genre and the action cinema.* London: Routledge.
Treichler, Paula, A. Cartwright, and Constance Penley (eds.). (1998). *The visible woman: Imaging technologies, gender, and science.* New York and London: New York University Press.
Webb, John. (1998). *Junk male: Reflections on Australian masculinity.* Sydney: Harper Collins.

Intersex: Beyond the Hidden A-Genders

Chris Somers and Felicity Haynes

Following Virginia Woolf, who believes that androgyny is where our "future salvation lies" (Weil, 1992:147), cultural feminists (Butler, 1990; Haraway, 1991; Garber, 1992; Lorber, 1994) have challenged the concept of gender categories as dual and oppositional. There has been growing awareness of the existence of gays and lesbians, transvestites and transsexuals who daily cross the barriers between male and female (Feinberg, 1996; Ekins and King, 1996).

Despite this, many persons who are aware that they do not neatly fit the binary have accommodated their mandated sex with discomfort and sometimes anguish. Those whose sexuality crosses the conventional male/female attraction are being tolerated more but only within bounds. Whereas we may accept, for example, the Sydney Gay and Lesbian Mardi Gras as a legitimate visual interpretation of sexual and gender preferences, we do so not only because it has become too loud and profitable a voice to be ignored but because the fascination with the extreme and eccentric performances that make up the Mardi Gras arises from our curiosity with the obscene, the display of something that was formerly hidden from view. The sad reality is that though these forms of expression may be acceptable on the day, they are considered taboo during the remainder of the year. Gays, lesbians, and bisexuals must as a necessity of survival keep within their own safe havens (Buchbinder, 1998) or at least be associated with the queer.

Some people defend transsexuality on the grounds that it has a "natural" basis, albeit a deficient one. Doctors Money and Ehrhardt, for instance, suggest that transsexuality may be a purely physical prenatal accident:

> The phyletic program may be altered by idiosyncrasies of personal history, such as the loss or gain of a chromosome during cell division, a deficiency or excess of maternal hormones, viral invasion, intrauterine trauma, nutritional deficiency or toxicity, and so forth. Other idiosyncratic

modifications may be added by the biographical events of birth. (quoted in Raymond, 1979:49)

Others reinforce the natural binary by classifying any gender-blender as a social transgressor, or deviant. Raymond herself, a former student of Mary Daly, believed that the basic indicator of sex is the chromosomal pattern of XX or XY and argued that transsexual women "are not women. They are *deviant males*" (Raymond, 1979:183).[1] She believed that the first cause of transsexualism was a patriarchal sex stereotyping system that led many transsexuals and women to feel hatred of their bodies and their functions (Raymond, 1979:175-176). Transsexuals, in Raymond's view, "contravene the laws of nature" by their androgyny and in their desire to appear "natural" become only a façade of that reality constructed through the advances in the prophylactic areas of pharmaceutical and surgical science. Surgeons, psychiatrists, psychologists, counselors, deportment instructors, speech therapists, and electrologists form powerful teams to fetishize forms, artifical vaginas and removed organs, reinforcing female stereotypes and keeping transsexuals in subordinate status. The principle of androgyny on which transsexualism rests, according to Raymond, is nothing but an unsatisfactory pastiche that adds up qualities thought to be masculine and feminine. She criticizes the legitimacy of male-to-female transgendered persons, sees these human beings only as males masquerading as females and dangerous to females. In so doing she not only fails to acknowledge the reality of the female to male, but ignores the reality of those who have a balanced physiognomy of male and female. Her ideology, formed within an exclusionary women's movement, does not allow awareness of the transsexual or androgynous experience (see Riddell, 1996:171-189). Ironically, in speaking against the reinforcement of stereotypical boundaries, she marginalizes those who wish to cross them.

Androgyny is not a new phenomenon. It has been present in fictional stories and artifacts for a long time. The growing number of personal accounts of lived androgyny may be seen to threaten a dominant patriarchy, and therefore androgynes are dismissed to the margins of queerness. But the growing public awareness of androgyny usually still relegates it to the status of a cultural phenomenon, a matter of choice and decision. In the general population there are a number of people who are naturally a sexual mix or poly-gendered. Intersex people still remain largely invisible, as Lee Anderson Brown points out. Intersex includes a wide range of people who are born with physiological characteristics of both male and female. In medical terminology there are people who are genetically male while presenting as phenotypically female and who are known as androgen insensitive; others who are complete mixtures of both; and others again who like Chris are termed pseudo-hermaphrodites or "aberrant males" who in the real analysis may well be more "female" than "male."

This paper makes more visible some of those people who identify physically as androgynous and are confident of their androgynous status. In medicine, particularly, intersex persons who are recognized as such at birth are normally reassigned as male or female by a complex formula drawn up by Harry Benjamin. Sometimes they are never told of their original status and their physical androgyny may be unknown even to themselves. Some of them may not be aware of their status as intersex persons, and yet they feel marginalized by their feelings of gender ambiguity. Their stories still remain very largely hidden and unknown.

Such a person is one of the authors of this chapter, Chris Somers. S/he is chromosomally 47XXY and therefore can be categorized medically as having Klinefelter Syndrome. Chris acknowledges that s/he has benefited in some respects from being medically categorized as having Klinefelter Syndrome, because without the testosterone prescribed to treat the accompanying osteoporosis, s/he might well be in a far worse condition than s/he currently is. But s/he has never considered him/herself as a deviant male. The characterization as "long-legged, with an undeveloped upper torso, mentally dull with poor social skills and retarded" (see "Introduction," 6) does not fit.

Despite his/her categorization at birth as a male, Chris always felt like a woman and at the age of nine was aware that s/he wanted to bear a child and have a neat vagina like his/her sister Louise. When at the age of seventeen his/her breasts began to develop, teachers and students at his/her private Catholic boys' school refused to value his/her differences. While s/he was proud of his/her androgynous body, s/he was teased mercilessly about it in the showers and his/her father, a doctor, advised him/her to have a double mastectomy. While s/he had some difficulty with intelligence tests at school, s/he is extremely creative, persistent, and verbally fluent, having won international acclaim for an international project in global education. His/her feminine interests were channeled into the creative arts, and respect for him/her was gained by excellence in sports, especially endurance sports like long distance cycling. S/he has been a guest speaker at conferences in Australia and overseas, has flown solo in a powered aircraft, and has illustrated several books for major international publishing houses. The stereotype of Klinefelter

Syndrome does not accommodate these exceptions. S/he was not labeled as such until s/he was 27, and many doctors still refuse to believe the X-ray evidence that s/he has a penis, ovaries, a collapsed internal vagina, and the possibility of a uterus, the latter discovered only at the age of 50.

Chris invites readers to consider the consequences of viewing themselves as a syndrome, perhaps called the "Somers Syndrome." Dreger (1998b:25) similarly says:

> I realized recently that I suffer from a genetic condition. Although I have not actually had my genome screened, all the anatomical signs of Double-X Syndrome are there. And while I could probably handle the myriad physiological disorders associated with my condition—bouts of pain and bleeding coming and going for decades, hair growth patterns that obviously differ from "normal" people's—the social downsides associated with it are troubling. Even since the passage of the Americans With Disabilities Act, people with Double-X remain more likely than others to live below the poverty line, more likely to be sexually assaulted, and are legally prohibited from marrying people with the same condition. Some potential parents have even screened fetuses and aborted those with Double-X in an effort to avert the tragic life the syndrome brings. Perhaps you know Double-X by its more common name: womanhood.

Though the difference between the labels 46XX and 46XY is minimal, the physiological differences between male and female can be astonishingly great. Similarly, all those who have different labels again, such as 45XO, 47XXY, 48XXXY and many others, have as much right to have their physiological and psychological differences recognized. The pathologizing of minority groups opens up for further research and investigation the whole issue of what constitutes sex and gender, especially when we realize that there is considerable variation even within one chromosomal category. To what extent gender and/or sexuality depend on chromosomes, and to what extent they are also modified by hormones, reproductive organs, external genitalia, body hair and shape, personal sense of self, emotional identity, or any combination of these remains largely unconsidered. We do not yet know if there are any other factors involved, such as the shape of the hypothalamus (Moir and Jessel, 1991; Angier, 1995).

The stereotyping that accompanies chromosomal categorizations is held by many to be as detrimental as beneficial to those so categorized. They become subject to the "gaze" of doctors and seen in terms of their pathologies rather than their capacity and diversities (Foucault, 1994). Chris illustrates the normalizing process within the medical institution with the following story. A 41-year-old woman who was 16 weeks pregnant was advised to have an amniocentesis. She was informed that her infant in utero was 47XXY (that is, had Klinefelter Syndrome) and given the option of terminating her pregnancy. Further she was advised she could be given drugs to induce complete amnesia of the fact that she had ever had a child aborted, as if she had come into hospital for some routine investigations. In this particular case the mother responded with confusion. She felt extremely depressed, confused, and suicidal at having to make such a decision. She felt that the obstetrician, a male, was more concerned about the possible mental retardation of the child than anything else, and appeared cold to her concerns and feelings, also telling her that the child would have major social problems in later life. While this might be true, no one had thought why that child might encounter those problems and that the problems might be caused more by an intolerance toward anything outside the "norm" than the "abnormality" itself. The cards are stacked up against the child being able to survive with dignity in a world with a narrow vision, even in a supposedly sophisticated society.

Chris knows of a similar case in New South Wales. Both parents had been pressured by geneticists and other members of their medical team to abort their 47XXY child as it would be socially inept, mentally retarded, and a poor representation of a male. They finally gave into pressure to have the fetus terminated. They confided their unease to a historian and former nurse (personal communication) who wrote:

> Someone at work recently had usual pregnancy tests (she's over 35), and was told she should have a termination as the foetus was XXY and this would result in severe mental retardation and psychological maladjustment. She knew I had worked in "psych" hospitals so she wanted to know what I'd seen of "these people," I told her I'd never known any while I was a nurse, but I do know a friend who is XXY who's recently returned to university to complete a Masters degree by research, and in the stunned silence added, & is no more psychologically maladjusted than you or me. The bottom line for her was that she & her husband couldn't face the prospect of having a child that was in any way different (which isn't necessarily predictable!) & didn't want to have a child who would probably have a difficult life, a natural parental wish, but not really predictable either. She's since had the termination, & was told quite insistently by the geneticist that XXY people are male but inadequately so. This is typical of the medical profession's patriarchal attitudes (I speak here as an ex-nurse), it's as if X's are just standard issue but Y's are special & absolutely defining. He could have just as logically said XXY people are female with a few extras, if he could only imagine two boxes to squash everyone into. This is apart from all issues involved with just how standardised do we really want or need our species to

be, & that would make for an unbearable life. The lady at work felt that by being different & not fitting in would be unbearable for her.

A sympathetic medical doctor commented that this was akin to genocide. She has a colleague and friend who was also advised to have her Down's Syndrome child terminated but refused and now has a healthy child whom she adores. What would have become of persons such as Stephen Hawking, one of the world's leading physicists, if doctors at the time could have predicted his future physical ailment? Would his mother have been advised to have him aborted before he could have made his own major contributions to the world?

The following narratives of intersex persons who have survived birth but have yet been continuously subjected either to severe normalization processes or treatment as freaks of nature bear further testament to their humanity and difference.

Sir Leonard Heron,* like Chris, was born genetically XXY and has had no "corrective" surgery. S/he has a beard and breasts, has a womb and ovaries, menstruates through the penis, and is technically capable of bearing a child by cesarean section, indeed longs to do so; s/he self-identifies as female but presents in daily life as male. At school s/he didn't participate in school sports, preferring to withdraw to the library and read books. The boys would call her/him "sissy" but this seems more related to her/his antipathy to sport and her/his preference for playing with the girls rather than to physical appearance, as Leonard was tall and not obviously physically different from others. Her/his parents encouraged her/him to always do the best she/he could and to endure suffering without complaint. Most primary teachers treated her/his inability to fit in with other high-spirited country boys as a weakness on her/his part and the first few years of school were desperately unhappy. S/he says s/he felt like a failure physically and loathed existence. Fortunately in fifth grade, a male teacher, who gave her/him special attention, noted her/his high intellectual achievements and pointed her/him to an academic career.

Currently, at the age of 53, her/his body causes him great physical discomfort, as s/he is in continual pain with osteoporosis and osteoarthritis, a condition associated with hormonal imbalance. S/he says, "I would have loved to have been born in 1830 or something like that. I feel more comfortable in the past," escaping the ugliness of modern life through books, film, and videos. S/he feels more like a female than a male, but is sexually attracted to females; s/he was married to a female with an adopted son. Having a male birth certificate, wearing his/her favorite silk shirts and a breast binder under a conservative grey suit, s/he was awarded a British knighthood, and s/he was also awarded the Military Cross for services in Vietnam. S/he has held a prominent position in the Queensland government, and s/he is articulate and widely read, currently teaching gifted children in a Queensland primary school. S/he is far from being inadequate and has probably succeeded in life beyond a number of the doctors who are advocating genocide for those who do not fit their criteria for "normalcy."

Leonard's isolation at school is echoed in other intersex stories. When asked what it was like for her/him in school and early life and what things were most significant to her/him in those memories, an intersex professional, Rhian, replied:

> I really don't remember much before about 10 years of age. I was a quiet kid and basically kept much to myself. I did a lot of competitive swimming right through my teenage years. This and study pretty much filled my life. I was very unhappy about living as a male and wanted to be a girl for as long as I can reliably remember. The swimming and study was just my way of coping with this distress. I lived in a conservative city and I had no idea that such a change was possible. I was very fearful of mentioning anything to my parents, etc. I went through a period of fairly bad behavior at school and the distress of it all got to a point where I had to make the decision: either I was going to swap over and live my life as a female or suicide, so I swapped over. Life, although not easy, has been so much better ever since. My reassignment gave me an inner happiness and peace that I never knew before. Surgery gave me a body which approximated that of a normal biological female and gave me the possibility of having a relatively normal relationship. Relationships, however, are always difficult. Because I want to be honest, any prospective partner needs to know my gender and intersex history. Most people cannot deal with it.
>
> Many men have a strong sexual curiosity, but as I am not a performing seal, I refuse to be associated with them and this type of scene. It is not the sort of relationship that I would find fulfilling. Some of the guys who show interest in me are simply gay and having problems dealing with their sexual orientation. They see me as some sort of intermediate and comfortable alternative. Well, I feel sorry for them, but I am a woman, not a guy. In any case, despite the social and legal difficulties, I have absolutely no regrets about undergoing reassignment at all. It has improved my life immeasurably and given me the possibility to lead a relatively normal, happy, and productive life: something that would not have happened without reassignment.

Rhian demonstrates the superficiality of Raymond's premise that the deliberate construction of a different gender is only artificial. There are many people who are very happy living as an assigned female, while they are aware that they were not naturally born so. As

Morgan Holmes, an intersex activist who has born a son, writes (in Feinberg, 1996:139):

> what many intersexed persons, myself included, have in common with women is that we also live in this world "as women." We were raised to be women, to be accorded only whatever rights and privileges women can manage to obtain, within the confines of [their] race and class in patriarchy. Because of the range of forms that intersexuality takes, there are some for whom it is easier to assume a single sex identity and some for whom it is more difficult, enforced through lifelong medical intervention.
>
> What is even more difficult than identifying oneself as a member of the community "woman" is attempting to define one's identity as an intersex/woman. The task requires taking back an identity which has been made illegitimate by culture and has been stolen through surgery.

Others are happy to present in daily life as male or female and maintain their "natural" body, without surgical reassignment. Glenda Lee* (46XYX) is an articulate and professionally qualified Australian Aboriginal, with a birth certificate that classifies her as a true female hermaphrodite. Despite living as an itinerant male truckdriver for many years, she bore a child *in utero* 25 years ago. She currently lives as a female hermaphrodite with a gentle male (46XY) partner. Her story is a mixture of abject abuse, racism, and alienation. "I was raped by both my adoptive father and brother at the age of 8 and 12 years of age respectively." This same person was repeatedly raped between the ages of 14 and 15 by the adoptive father and a friend of his and later ran away from her oppressive "home" after being severely beaten by her father for asking why she was subjected to such brutality. She is not bitter, but has said, "I have never had a life and have been to hell and back and beyond."

Glenda was brought up as a boy and taunted at school. She was forced to use the male toilets where she sat down to pee with her peers looking under the door, climbing over the top of the partitions, all hoping to have a look at the "curious show that was amongst them." She left school early, hitch-hiking from state to state, saying: "I became a street kid and was living from the pickings from garbage bins, sleeping in cardboard boxes and lived in constant fear of the world as a dejected and abused young person." Presenting as a male, she worked at casual labor for many years. When later she decided she wanted to live as a female, she trained as a psychiatric nurse, and worked in a gender clinic giving advice to those who suffered from gender dysphoria, marrying a female to male transsexual for a short time. Glenda's life has been a nightmare of appalling atrocity where the authorities had interest in her only as a subject of hushed medical curiosity. She had to suffer the indifference of the police force to genuine pleas for help after being raped again as recently as three years ago. It is a tribute to her positive thinking that she has sought to help others with gender identity disorder.

Janet Marteene,* born in 1940 as 46XX but with ambiguous genitalia, was not given a birth certificate at the time, but nearly 20 years later obtained a declaratory certificate which said she was born on 16 August 1959. At chronological age 5 she frequently enjoyed pulling apart electrical objects and reassembling them, which was then considered an inappropriate pastime for girls. At 10, without explanation or counseling, she was whisked into hospital and assigned a male gender. A shunt was placed in her vagina, which was sewn up. She was placed on massive doses of testosterone, which she says contributed to later psychoses. As soon as she could read and write, she stopped attending school and studied and worked at home repairing and building electrical appliances for friends and acquaintances. She had a natural aptitude for advanced math and sciences and taught herself at home from books including the Miniwatt technical data book. Her IQ results "go off the scale," with a score of over 200. She obtained a senior commercial pilot's licence, flew jet ranger helicopters and was the aerobatics pilot of a Pitt Special aircraft. She is now an electronic engineer, computer scientist, and computer artist working from home, and having decided that she is really female, now lives as a woman with her female partner after years of botched surgery and inappropriate hormone replacement therapy. She lived in pain and seclusion, on the brink of psychotic episodes for many years until she saw Chris appear on a *60 Minutes* program with Jeff McMullen on the Channel Nine TV Network in 1995. She said that the program "gave her hope." Her story of mismanaged medical intervention parallels that of John/Joan (Colapinto, 2000) except that John managed to avoid surgical reassignment and has resumed his perceived identity as a male.

All of these people have at some stage denied aspects of their humanity to meet public perceptions of the norm, aware of the physical or vocational dangers of being too obviously queer or different. From the extreme process of eugenics, through surgery, through normalizing vocational expectations to a relatively slight matter of requiring gender-specific dress, society imposes its norms in many ways. Education has as one of its aims the normalizing process, but this can often come into violent conflict with its paramount aim: to foster the development of each child as an autonomous and unique being. We do not know how many other suppressions have occurred. We do not know how many people have never revealed their androgynous physicality to anyone, or how many have never been made aware that their gender has been assigned to them by doctors rather than by nature. We do not know how many have been rendered invisible and in that sense dispossessed of their identity.

Toinette M. Eugine (1996:461) spoke of a similar dispossession of ethnic minorities:

> As long as we feel insecure as human beings about our bodies, we will very likely be anxious or hostile about other body-persons obviously racially or sexually different from our own embodied selves. Thus, the most dehumanising spoken expressions of hostility or overt violence within racist and/or sexist experiences are often linked with depreciating the body or body functions of someone else. Worse yet, though, the greatest dehumanising or violence that actually can occur in racist and/or sexist situations happens when persons of the rejected or racial- or gender-specific group begin to internalise the judgements made to others and become convinced of their own personal inferiority. Obviously, the most affected and thus dehumanised victims of this experience are Black women.

Androgyny was for many romantics of the 1970s (Eliade, 1969; Heilbrun, 1973; Singer, 1976; Ornstein, 1973; Watts and Elisofon, 1975) a mythical ideal, depicted in artworks of various cultures as a place where animus and anima combine in one body, where the yin and the yang come together as one. We have tried to present to you the stories of some people for whom androgyny is an actuality and for whom even this "third gender" takes various forms. The physical realities are seen by many to be threatening because they render fragile some deep-seated assumptions about humanity. The truths of androgyny are, however, too complex even to be categorized easily in stereotypical syndromes. Many doctors make the stereotypes true by forcing the "patient" into the normalizing shape of male or female, making the binary true by surgery or genocide. So do many teachers, without realizing it. Our assumptions concerning a bipolar social construct in Western society are extremely naive. The reality is opposite to the popular belief that to be a person one must be either male or female. We must recognize that gender and sexuality is a complex continuum to be cared for in the mystery of life.

If people, especially children, are to be given the opportunity of defining and realizing their own identity, then teachers and those empowered to help with the process of becoming have to learn that they may regard something as "freakish" because of their own ignorances and linguistic constraints. Knowledge empowers. We could rid ourselves of the fear of falling off the edge of an assumed "flat earth" only by learning that there were no edges to fall off. Similarly we can accept a person being both male and female when we learn that the boundaries of sex and gender are more fluid than we have been taught to believe. Only if teachers and students can recognize the person beneath the surface features of sex and gender can a humane education that recognizes and values diversity take place.

References

Angier, Natalie. (1995). Size of region of brain may hold a crucial clue to transsexuality, a study finds. *New York Times*, 2 November, A12.

Benjamin, J. (1997). Sex reassignment at birth: Longterm review and clinical implications. *Archives of Pediatrics and Adolescent Medicine*, 151:1062–1063.

Buchbinder, D. (1998). Homosexuality and the diasphora. Unpublished paper presented at Postmodernism in Practice Conference, Adelaide, February 1998.

Butler, Judith. (1990). *Gender trouble: Feminism and the subversion of identity.* New York: Routledge.

Colapinto, John. (2000). *As nature made him: The boy who was raised as a girl.* London: Harper Collins.

Dreger, Alice. (1998a). *Hermaphrodites and the medical invention of sex.* Cambridge, MA: Harvard University Press.

Dreger, Alice Domurat. (1998b). Ambiguous sex—or ambivalent medicine? Ethical issues in the medical treatment of intersexuality. *Hastings Center Report*, 28 (3):24–35.

Ekins, Richard, and Dave King (eds.). (1996), *Blending genders: Social aspects of cross-dressing and sex-changing.* London and New York: Routledge.

Eliade, Mircea. (1969). *The two and one.* (Original title *Mephistopheles and the androgyne).* New York: Harper Torchbook.

Eugine, Toinette M. (1996). While love is unfashionable: Ethical implications of black spirituality and sexuality. In A. Garry and M. Pearsall (eds.), *Women, knowledge and reality.* London and New York: Routledge.

Feinberg, Leslie. (1996). *Transgender warriors.* Boston: Beacon Press.

Foucault, Michel. (1994). *The birth of the clinic.* Trans. A. M. Sheridan Smith. New York: Vintage Books.

Garber, Marjorie. (1992). *Vested interests: Cross-dressing and cultural anxiety.* New York and London: Routledge.

Garry, A., and M. Pearsall (eds.). (1996). *Women, knowledge and reality.* London and New York: Routledge.

Gasche, Rodolphe. (1995). *Inventions of difference—On Jacques Derrida.* Cambridge, MA: Harvard University Press.

Haraway, Donna. (1991). *Simians, cyborgs and women: The reinvention of nature.* New York and London: Routledge.

Heilbrun, Carolyn. (1973). *Towards androgyny: Aspects of male and female in literature.* London: Victor Gollancz.

Klinefelter, H. F. (1986). Klinefelter's Syndrome: Historical background and development. *Southern Medical Journal*, 79 (90):1089–1093.

Lancaster, Roger N., and Micaela di Leonardo (eds.). (1997). *The gender sexuality reader—Culture, history and political economy.* New York and London: Routledge.

LeShan, Lawrence. (1976). *Alternate realities: The search for the full human being.* London: Sheldon Press.

Lorber, Judith. (1994). *Paradoxes of gender.* New Haven, CT, and London: Yale University Press.

Moir, A., and D. Jessel. (1991). *Brain sex: The real difference between men and women.* London: Mandarin Paperbacks.

Ornstein, Robert. (1973). *The nature of human consciousness.* San Francisco: W. H. Freeman.

Raymond, Janice. (1979, reprinted 1994). *The transsexual empire.* Boston: Beacon Press.

Riddell, Carol. (1996). Divided sisterhood: The transsexual empire. In Richard Ekins and Dave King (eds.), *Blending genders: Social aspects of cross-dressing and sex-changing.* London and New York: Routledge.

Schwartz, Adrian E. (1998). *Sexual subjects—Lesbians, gender, and psycho-analysis.* New York and London: Routledge.

Singer, June. (1976). *Androgyny.* New York: Anchor Press.

Watts, Alan, and Eliot Elisofon. (1975). *Tao: The watercourse way.* New York: Pantheon.

Weil, Kari. (1992). *Androgyny and the denial of difference.* Charlottesville and London: University Press of Virginia.

Woodward, Kathryn. (1997). *Identity and difference—Culture, media and identities.* London, Thousand Oaks, and New Delhi: Sage Publications, in association with The Open University.

Notes

* All interviews were conducted in confidence and anonymity guaranteed unless consent was given. Leonard Heron, Janet Marteene, and Glenda Lee allowed their real names to be used in association with their intersex status for the first time at the Third International Congress on Sex and Gender, at Oxford University in September 1998. It is a testament to their courage and the growing awareness of diversity. None should be afraid to step "out of the shadows," walk tall and be recognized as equal citizens.

1. Fifteen years after its publication, after Johns Hopkins had dismantled its Gender Identity Clinic, Janice Raymond (1994:xi–xxxv) reissued *The Transsexual Empire* with a new introduction reiterating her views that transsexual surgery is the invention of men, initially developed for men, and that men cannot be "real women." In the introduction she made more explicit her radical feminist agenda that all gender conformism should be politicized and transcended.

A Sometime Woman: Gender Choice and Cross-Socialization

Michael A. "Miqqi Alicia" Gilbert

Introduction
Social construction theory holds that social and cultural institutions are created through a historical and social process that reifies custom and practice to the point at which it has the appearance of natural, unalterable fact. Contemporary feminists have used this view to argue that the differences apparent between the sexes in modern society are not necessary biological facts, but contingent matters arising from deeply embedded traditions and practices. As such they can be altered through appropriate means to allow less sexist, and more egalitarian relations between the sexes.

Contemporary gender theorists go further. They have applied social construction theory to the binary gender system, arguing that the very division of the world into two distinct genders correlated with two distinct sexes is an artifact of social construction and can and ought to be modified. Judith Butler and Marjorie Garber, for instance, view the gender rule-breakers as proof that gender is a variable and not a constant, that one can change one's gender, construct one's sex, or maintain a status not directly identifiable as that of one or the other classic gender.

Most of the discussions on this subject have focused on two main groups. The first consists of transsexuals who move entirely from one gender/sex locus to the other, and the second consists of drag queens and kings who, by and large, identify as their birth-designated sex, but who play at or perform as the opposite gender. More or less lost in this formula is the run-of-the-mill cross-dresser who periodically adorns himself in women's clothes and participates in activities ranging from hiding timidly in a drapes-drawn home to marching boldly into shops and restaurants.

In this essay I explore the relationship between the idea of social construction, its main tool, socialization, and the difference between "being" or "feeling" and "imagining." I argue that the public nature of socialization allows any cross-gendered individual the possibility of becoming a member of the "opposite" sex, but that the average cross-dresser does not have a strong or early enough identification to qualify on this ground. On the other hand, there is no reason why an individual, based on accrued public socialization available, cannot sometimes explore and enter certain of the constructs applicable to a non-birth-designated gender.

Assumptions

The idea of gender as a social construction has played a very important role in recent gender theory. Gender roles are seen as social, political, and institutional constructs. The essential axiom is that our understanding of what it is to be a man or a woman, male or female, is not something innate and unchangeable, but rather something that is created by social and historical forces.

The implications of such a view are profound, not least for the transgender community. In general terms such a view means that gender roles are open to modification with an eye to equality, the elimination of oppression, and the lifting of the barriers and expectations connected to each gender. If gender is a social construction, then gender roles are similarly creations of sociohistorical processes. That means that such historical realities as the types of employment and the educational, domestic, and economic opportunities that are possible can be changed.

It is also the case that the social construction of gender does not mean it is in any way uniform across cultures, regions, or nationalities. Just because it can be viewed as an actuality developing out of a collective sociohistorical process, it makes sense to view it as distinct within different areas. No one thinks for a moment that being a Muslim woman in rural Iran is the same as being a woman who was born and raised as an urban New Yorker. These two women may have, in fact, very little in common with each other aside from the fact that they are both in a society where being a woman is valued less than a man. So, the idea of the social construction of gender does not mean that gender roles among differing regions and peoples will be the same.

Taking this obvious theorem one step further, we develop a corollary. Consider two people, let's call them Eve and Louise, who grew up in two different families in the same geographic region. On the one hand they would both be exposed to the same general institutional socialization processes such as school, church, and the general societal values communicated through the media. But we might also expect differences. Eve might have gone to a public school, while Louise attended a religious school. Eve might have been raised by a single mother with strong feminist feelings, while Louise was raised in a traditional family with a male-dominated hierarchy. Eve might have been a single child, while Louise was the youngest of five with two older sisters and two older brothers.

Obviously, the ways in which Eve and Louise perceive gender, gender roles, and the flexibility and variation within them are very likely to differ, perhaps markedly. On the other hand, the two might have become best friends and had a tremendous impact on each other, thereby diminishing their differences. Or, to complicate matters further, we can introduce Sally, Louise's older sister, who, unlike Louise, vehemently rejected every gender role assumption that other family members embraced. In other words, there is no way to tell what values a given individual will hold or what perceptions will color her worldview. The simple fact of social constructionism does not mean that people in similar environments will (or will not) share values or absorb the same socialization (Chodorow, 1995:516–544).

To summarize, then, I am assuming that some level of the social construction model obtains, but I am further assuming that individuals end up with a gender image and ideology that is unique. Just as we can readily accept that social forces differ according to geographic and cultural variables, so it should be clear that the microcosm in which each person develops must have its own unique climate.

Socialization

It is time for the nitty-gritty. How does all this impact on the transgendered individual? Obviously, the impact of any theory that claims that physical factors, notably genetics and somatic configuration, are not the key or sole determinants of gender will be very meaningful to the transgendered individual trying to figure out why she is who she is. Most interestingly from the point of view of this discussion, as we track an individual transgender life we can see vital differences between individuals and categories. The closer we look, the more it becomes clear that every individual defines a unique gender niche, and only when we generalize in such a way that the individual is lost do we come to categories.

Consider, for example, a scenario recognizable to many crossdressers. It centers on a young lad who, when maybe 7 or 8, was not good at sports, was sensitive, cried easily, was fearful of physical injury, but was "cute" and clever. This little boy, let's call him Michael, looked around and observed that the qualities he happened to possess were valued or acceptable in girls, but not in boys. Wistfully, he decided he would make a better girl than boy, and wished he could be a girl and be judged by girl rules rather than boy rules. Michael did not think that he was "really" a girl and not a boy. He did not believe that some terrible mistake had been wrought; nor was he initially fascinated by clothing. On the contrary, what attracted him was the idea of the *social role*, of the difference in responsibilities and privileges as witnessed by him. What is important is the great degree of *choice* involved here. But choice is always involved. To Michael's young mind, unschooled in the pitfalls, hazards, and demons to which young girls are subject, there was no contest: girls had it better. He might well spend a great deal of time observing girls, being with girls, and absorbing, within a fantastical milieu, the socialization intended for them. He *chose,* consciously, to examine the rituals of the "opposite" gender.

At the opposite extreme, many accounts of cross-gender identification, specifically the accounts of many transsexuals, emphasize that the individual in question has *not* made a choice, has not *decided* that she is not a girl and is really a boy or would prefer to be a girl rather than a boy. To the contrary, she describes the situation as one in which the correct internal gender identity was simply always there. So, for many people it is the *lack of choice* that is the most compelling aspect. But at the same time, the subject does, in fact, constantly choose socializing priorities and influences. While the compulsion to cross-identify may, phenomenologically, mask the constant choosing, it is there nonetheless. It is there every bit as much as in a non-cross-identified individual who constantly chooses, filters, and accepts bits and streams of socializing information.

And here we begin to notice possibilities for differences in socialization. The more one identifies as a member of a gender, the more the socializing forces of that gender will impact on one. Michael's choice was to compartmentalize his gender structure. He wanted to pick and choose: "I'll have two from gender A and three from gender B, thank you very much." In the beginning there was no fetishization or eroticization of clothing at all. Later, this somehow segued into a fixation for clothing, which only later re-evolved into a gender identification issue. But someone who chooses the non-birth-designated gender, whose identification is complete, is open to the myriad socializing influences available. Socialization directives and instructions are out there, like radio waves, and all one need do is to tune into the preferred station.

There are, however, major stumbling blocks. The first difficulty, regardless of how strong the identification is, is the inability to practice. The transgendered person, and especially the male to female (MtF), is typically denied the right to act out the socializing instructions he is absorbing. The female to male (FtM), under the guise of tomboy-ism, can indulge in some practices, though even here real rebellion, (e.g., refusal to "dress up pretty for Aunt Agatha") can lead to censure. So the little boy who identifies as a girl and has learned to keep quiet about it seems to be socializing as a boy. But how is this known? The clothing he wears, the toys he plays with, often his companions and playmates, are chosen for him by his parents and various social institutions. The concentration on socializing motifs, though, is up to him. Beneath the hood, as it were, he is letting himself be influenced by whatever social messages *he* chooses to identify with; he is choosing his gender in choosing his socialization. In other words, a young *Homo sapiens* with a vagina might watch TV and, when seeing a pretty little girl happily cavorting in a party dress, say "Yuck," and reject the idea of identifying with this image. Similarly, a young human with a penis might have a strong identification with that same feminine image. But the limitation, for all transgendered individuals (TGs) is, invariably, the need to hide the cross-gender attachment from parents, friends, and teachers. This means one cannot indulge in the day-to-day living, the habituation, that is the essence of gender. But it does not mean that the socialization is not taking place on a hidden level.

We must never forget, in all of this, how early on most TG folk realize that when they cross-identify they are doing something wrong, something that must be kept hidden and secret. Little boys learn that if they act sissy-ish, they will be ridiculed and beaten. Little girls learn not to announce that they are not "tomboys" but real boys. This usually results in a kind of underground thinking that will, in time, develop into a full-blown gender outlaw personality.

The second difficulty is that a youthful gender offender is not able to catch all of the socializing influences available to a "natural" member of a gender group. It is extremely unusual for a child who chooses a divergent gender to be permitted to follow through on that identity. Various sorts of socialization will not be available or easily accessible. Most notably, the interactions that typically take place wholly within one gender group are not available to outsiders. Thus, the sort of intense socializing that takes place among a group of girls when there are no boys around is the sort of experience largely unavailable to the gender offender. Similarly, the interactions between a mother and daughter, between a female grade school teacher and female students, and so on are experiences from which the birth-designated boy is largely excluded. The occasional tomboy might experience limited acceptance outside of sports situations, but will also quickly be identified as a girl when many classically male discussions begin. One of the strongest socializing forces is the abundance of people who quickly point out what gender you are supposed to be, and that you do not belong somewhere simply because you want to.

The earlier and the stronger that cross-gender identity is formed and consciously recognized, the more likely it is that one can subvert the mainstream forces and covertly cross-socialize. There are so many variables that it is impossible to generalize. But the biggest gap for the cross-gender child must lie in the area of intragender activities. He is excluded from learning, among other things, the modes of communication used and nexus of values held by his cross-gender peers. He may see instances of these activities portrayed in the media but the intricacies of activities, the subtleties and nuances of what is right, what is wrong, how to behave, what to believe, which attitudes are valued and which denigrated come to the young person very largely through the medium of interaction.

Social interaction for the young continually involves correction, change, and alteration. Young girls constantly correct each other's fashion sense—"You're not wearing *that*, are you?" They forever pass around the lore of nature, the wisdom of women that has been handed on to them. Women are not born knowing what "matches," but they always do know, and most men do not. Similarly, men have no innate nature that causes them to pause and let a women precede them through doors. That is something trained into you when you are quite young, and conditioned until (certainly for my generation) it becomes second nature. As a man, you must learn what to laugh at, how to be lewd, and how to participate in the kind of competitiveness that establishes the hierarchy existing within every group of boys.

Socialization is the most crucial component of gender identity, and the most effective method our culture has for maintaining the duality of sex and its one-to-one identification with gender. It is precisely why gender offenders must be treated so harshly, why they must be humiliated, berated, scolded, punished, admonished, and even beaten. Not to do so is to weaken the bipolar gender system, and that would undermine the entire structure on which our patriarchal system depends. If you cannot tell who's a boy and who's a girl, then how do you know who is higher or lower on the power ladder? How do you know who can be exploited, who can be raped? How do you know who is supposed to be the doctor and who the nurse? Without the clarity of the bipolar system, chaos would reign.

Being, Feeling, and Imagining

Let me return now to the beginning. The transsexual earns her stripes by an early cross-gender identification resulting in an awareness and absorption of non-birth-designated socialization messages. There are limitations to this socialization, but the stronger the identification, the more likely the limitations are to be overcome. Clandestine socialization may not be as far-reaching and thorough as aboveboard socialization, but it will work. For one thing, the idea that there is only one way to fulfill the social constructs that define woman-ness is, as we saw above, naïve. It is easy to find persons defined and accepted as women who are worlds apart in terms of outlook, perspective, behavior, and values. The transsexual's socialization might mean that he becomes a man who is somewhat different from a stereotypical, mainstream, traditional male, but then so are many "natural" males.

So, one question is, as far as I am concerned, answered in the positive: Can a male transsexual *be* a woman? Can a female transsexual *be* a man? Yes. Why? First, because socialization is public and available to those who choose it. Secondly, because not every aspect of the classical socialization is necessary for an individual to *be* a member of that gender. A genetically born and self-identified woman raised in a situation where she had little contact with other women would still be unquestionably a woman, even though she might have missed some of those very intragender experiences missed by the transsexual.

What then of the cross-dresser? Could I, as a result of my cross-dressing, *be* a woman? I know that I am not actually a woman, so can I be a *sometime* woman? Can a cross-dresser step in and out of a gender? Is gender something that can be taken on or off, like a suit of clothes? (Pun intended.) Consider the method actor who says, "When I play a cowboy, I must *become* a cowboy, I must feel like a cowboy in every fiber of my body." But, of course, the actor really knows little about being a cowboy. He has picked up some bits of information here and there from the media, perhaps read some books, some biographies, but he has, nonetheless, the sense that he *feels what he imagines a cowboy feels*. The great actors, we are told, *become* the characters they portray. But, then, Erving Goffman (1969) tells us that we act out who we want to be, that we are all great actors, and that, indeed, "all the world's a stage."

There are numerous differences between the cross-dresser and the transsexual, but one is critical and, perhaps, defining. For the cross-dresser there are often marked variations in feeling, style, attitude, comportment, and voice. The cross-dresser is not, typically, in one mode all the time. Like the actor who can go home and be himself, the cross-dresser moves in and out of the role, in and out of the feeling. But, because the internal mental state is changeable, it does not make sense to say that the cross-dresser is a woman (or man). The theory of the social construction of gender does not permit the cross-dresser to *be* a sometime woman, but it does allow the cross-dresser to *imagine* that he is a woman and even to *feel* womanly. How can this be? How can a cross-dresser *feel* like a woman even though he makes no pretense of being one? How can a cross-dresser be manly even though she does not believe herself to be male?

The social construction of gender relies on there being certain standard characteristics inculcated into the individual from birth. Some of these are highly internal and difficult to define, for example, communicative characteristics. Others are much more public and more easily recognized, for example, clothing and mannerisms. Now, it is vital to understand that multitudes of easily identified women violate many or most or perhaps even all of the rules or characteristics that social mores ascribe to their sex. This is extremely important from the point of view of feminist theory and understanding gender. However, for the cross-dresser the visibilities provide the means of access. They provide a means whereby the cross-dresser creates the image and draws forth the feeling of woman-ness. Like the actor who chooses the visible characteristics of a cowboy in order to conjure forth the feelings such costumes and mannerisms generate, the cross-dresser uses the artifacts of clothing, posture, and mannerism to become womanly. The actor is not, *ceteris paribus*, going to take as a role model a cowboy who wears a three-piece suit. Similarly, the female cross-dresser is not going to emulate a man who is effeminate, and the male cross-dresser is not going to choose a butch lesbian as a model.

Typically, what the cross-dresser has access to is *femininity*. And, at that, femininity as publicly defined and identified via the various media, observable behavior, and cultural institutions. It is femininity, not some internal sense of "woman," that is the most public aspect of woman-ness, and one that is most adoptable. In other words, the cross-dresser cannot be a *woman* but can be *feminine*. And, through the adoption of femininity the cross-dresser is able, with insight and care, to use that as an opening, as a window to explore what it means to be a woman. Similarly, the female cross-dresser who does not imagine herself to be a man, but dresses in a masculine way, affects a swagger, takes up street space the way a man does, and so on, can imagine, can feel what that means, what its implications are, what differences it makes for one's existence.

The theory of social construction does not permit a cross-dresser to be a woman, because there is no such thing as a sometime woman. Whether one is birth-designated a woman or not, one can be a woman, but not sometimes. On the other hand, the very ideology on which the theory of social construction rests means that there are loosely identifiable and occasionally mimic-able phenomena that are gender-specific. The adoption of these artifacts and attitudes can induce a feeling that one imagines is woman-ly. This feeling can, in turn, receive positive and negative feedback in various circumstances.

The publicity of socialization permits us to know what to do and how to act in numerous circumstances. The social construction of gender places an enormous weight on that socialization, and, at the same time, makes it available as something that can be borrowed. We all, in this community, know the great gulfs that exist between some people's self-image and their reality. I know male cross-dressers whose idea of femininity begins and ends with garter belts, and female cross-dressers whose only idea of being male is carrying a pack of Marlboros rolled up in a T-shirt sleeve. But there are a multitude of others who, regardless of where they begin, go well beyond such facile images. Just as the little boy who was not good at sports came to realize that girls had it as hard as boys, so cross-dressers who are permitted to grow, learn, and explore come to learn far more about being a woman, about being a man, than most people are willing to allow.

At this point we move into the realm of the quality of the experience rather than its existence, and there is no time for that. Clearly, there is more to be said here about the effects of experience and age, socialization within cross-dressing communities, the integration of personae, and the depths of the selected gender one can plumb. But these further musings must wait for another platform. For now, it suffices to answer the question implied in the title of this chapter: social construction has limits, and one of those limits is that one cannot *be* a sometime woman.

References

While I have made very few direct references in the text, the following works are among those that have made particular contributions to my thinking on this subject.

Bem, Sandra Lipsitz. (1993). *The lenses of gender.* New Haven, CT: Yale University Press.
Bornstein, Kate. (1994). *Gender outlaw: On men, women and the rest of us.* New York: Routledge.
Bullough, Bonnie, and Vern Bullough. (1997). *Gender blending.* Buffalo, NY: Prometheus.
Bullough, Vern, and Bonnie Bullough. (1993). *Cross dressing, sex, and gender.* Philadelphia: University of Pennsylvania Press.

Butler, Judith. (1987). Variations on sex and gender. In Benhabib, Seyla, and Drucilla Cornell, eds. *Feminism as critique*. Minneapolis: University of Minnesota Press.

Butler, Judith. (1993). *Bodies that matter: On the discursive limits of "sex."* London: Routledge.

Chodorow, Nancy J. (1995). Gender as a personal and cultural construction. *Signs*, 20:1:516–544.

Devor, Holly. (1989). *Gender blending: Confronting the limits of duality*. Bloomington: Indiana University Press.

Feinberg, Leslie. (1993). *Stone Butch blues*. Ithaca, NY: Firebrand Books.

Garber, Marjorie. (1989). Spare parts: The surgical construction of gender. *differences: A Journal of Feminist Cultural Studies*, 1:3:137–159.

Garber, Marjorie. (1992). *Vested interests: Cross-dressing and cultural anxiety*. New York: Routledge.

Gilbert, Michael A. (1997). Beyond appearances: Gendered rationality and the transgendered. In Bullough, Bonnie, and Vern Bullough. *Gender blending*. Buffalo, NY: Prometheus.

Goffman, Erving. (1969). *The presentation of self in everyday life*. London: Allen Lane.

MacKenzie, Gordene Olga. (1994). *Transgender nation*. Bowling Green, OH: Bowling Green State University Popular Press.

Vance, Carole S. (1995). Social construction theory and sexuality. In Berger, Maurice, Brian Wallis, and Simon Watson, eds., *Constructing masculinity*. New York: Routledge.

Wittig, Monique. (1981). One is not born a woman. In Wittig, M., *The straight mind*. Boston: Beacon Press.

Making a Transgenderist: The Construction of Gender Identity in Boston and Amsterdam

Adrianne Dana-Tabet

Ethnographic investigations in the Boston area and in Amsterdam indicate that significant differences exist in the definition and construction of Dutch and American marginal gender and sexual categories. For instance, the expandable umbrella category of transgenderism, so prevalent in gender discourses in the United States, was introduced in Holland only two years ago. Its unique meaning in Holland relates to the construction of a still relatively insignificant but specific gender identity category, the transgenderist, with discrete criteria and a developing protocol for medical interventions and legal validation. In general, however, under the auspices of both the Dutch legal and medical establishments, gender and sexually *confused* individuals are either oriented toward transsexualism and eventual government-supported sexual reassignment surgery or relegated to the unsubsidized, socially marginalized category of fetishistic crossdressers. In this case, government facilitation, contrary to popular conceptions of Dutch liberalism and tolerance, may actually contribute to the perpetuation of the traditional sex and gender distinctions of male and female, as individuals configure identity and behavior to conform to these binary classifications. This kind of control differs significantly from the sanctioned proliferation of sexualities in Holland, which is reminiscent of Foucauldian conceptions of state power. In contrast, transgendered individuals in the United States, while limited legislatively in achieving essential rights and protection, have the opportunity to embrace a range of gender and sexual identities while creating a plethora of discourses that reflect mainstream assimilation or *outlaw* perspectives.

The term *transgenderism* continues to undergo considerable transformations and cross-cultural translations as the U.S./Dutch cases suggest. Originally coined in the U.S.A. by Virginia Prince, a pioneer in alternative gender ideology, *transgender* as a gender category, identity, and repertoire of behaviors was initially introduced in the 1950s in Prince's seminal publication, *The Seahorse*. Recently, at the International Foundation for Gender Education (IFGE) conference in Toronto, Virginia commented on the origins of the term. She said:

> I suppose it's my own myth making. Though I had been a crossdresser for many years, I began dressing full-time and was no longer a transvestite on an episodic basis. I changed my gender and therefore needed a noun. I'm a transgenderist because I wanted a term that would give me a handle, a name like Virginia. I also wanted to make a distinction between what I was doing and people wanting surgery—to give them something else. It's like this —you can fly from New York to L.A., and I got off in Chicago.

Categories by their very nature, seem more a function of analytical purpose than a reflection of the lived experience of participants. By maintaining a discrete transgender category, which specifically excludes motivations and behaviors that might be construed as fetishistic or exhibitionistic—which classifies individuals as cross-dressers; or gender dysphoric—which places individuals on the transsexual track, the reality of a fluid gendered identity transformation is ignored and negated. As the ethnographic evidence indicates (and for the purposes of this discussion only the remarks of male to female transgendered persons have been included), life constraints, as well as socialization processes, transform the individual's perception of the gendered self, facilitate the exploration of various sexual orientations, and invite the reconceptualizations of alternative gender roles.

As Carla/Rob, a Boston informant, reports:

> It is important for me to find out exactly where I fit on the transgender spectrum. This is an issue I'm beginning to think and talk about alot. For me dressing was a *sexual thrill* especially in childhood and youth. Now, though it's still a pleasurable and sensual experience, it's become something else—a compulsion that relieves anxiety, yes, but an expression of myself or another part of the self. I'm not sure if I am a transsexual and at this point my goal will be to live part-time as Carla. If Carla were out all the time, I would miss Rob or aspects of Rob.

During the past eight months, Carla has come to consider herself to be transsexual. She lives full-time as a woman and has begun the difficult process of transitioning at work. Currently engaged in the psycho-therapy and hormone replacement therapy necessary prior to sex reassignment surgery (SRS), she remains uncertain about committing to vaginoplasty. What is evident in this brief example is

Making a Transgenderist 53

Carla's desire or need to locate an identity and the process by which variations in behavior and the meaning attributed to practices change over time. Carla has moved from the discourse of prurient erotic experience to one of self-discovery and actualization.

Kimmi, another Boston informant, explains:

> I consider myself to be transgendered, in the umbrella sense of the word. I found that calling myself either TS or CD [transvestite, cross-dresser] implies an either/or type of identity rather than the both/and that I consider myself to be. I will never be a girl. I am and will always be transgendered. I have large hands and feet. My hair and breasts go back in boxes. Initially I wasn't okay with this realization and I had to get there. I think I've developed a different way of handling the dysphoria. One day I figured out that I'm simply not going to forsake or deny an extremely strong feminine component in me, but—I'm not ever going to be female. I admit I've considered castration, but only because I hate the masculine chemicals—they make you territorial, aggressive, and dumb.

Kimmi is speaking the language of inclusion and exclusion simultaneously. While embracing a variety of gender identities, she uses the term transgender to locate herself or her own identity as a panoply of gender permutations. In a sense, she defines herself by what she is not—not a cross-dresser, not transsexual, not a female, not a male.

Brenda, an active member of the Boston transgender community, states:

> I consider myself transgendered because the term defines anyone with a fascination for clothing or whatever that is not the gender norm. My cross-dressing goes beyond fetishistic things. In the beginning, it was a sexual release and I used it specifically for masturbatory purposes. There was a nightly ritual where I dressed in nightgowns, masturbated, and felt very guilty. I remember fantasizing about being a woman and prayed to wake up a woman. I also had a fascination with breasts and wanted my own. I still have a fetish with breasts and nipples. We all come into the TG community as we come into the world—innocent and vulnerable. We know nothing about cross-dressing or TS issues and everything in between. I don't know where I'll be 2 years from now. It goes beyond being transgendered to just trying to figure out the next part of my life. As far as becoming a woman, I would not want to be a woman every day. I'd like to integrate parts of it into my life, but the transformation is too all-consuming. I really don't want to lose my penis and I feel as though I've found a comfort level. I can expose that repressed part of my personality.

Brenda's conception of transgenderism combines the classical definition of cross-dressing—an eroticized fascination with the clothing of the opposite gender—with an ideological awareness of transgressing gender norms. Hinting at the process of community enculturation, Brenda conjures images of a *tabula rasa* that requires that the uninitiated—the "innocent and vulnerable"—embark on a journey in search of experiences that shape gendered perceptions of self and dictate courses of action. Barbara has continued to opt for an integration of male and female attributes without choosing to undergo an "all-consuming transformation." Recently, she has told me that her need to dress en fem for social purposes has decreased, although the experience is just as pleasurable as in the past. Her more androgynous choice of attire seems for her more reflective of the internal integration she has been experiencing throughout her alternative gender explorations.

In Holland, issues of gender identity and concomitant behaviors are complicated by medical protocols, health insurance reimbursement, and Dutch family law. The struggle to create a legitimate discourse of transgenderism and a transgenderist identity has been led by a few rebels who have refused the validation offered to transsexuals willing to embark on the process of gender/sex reassignment or to accept the stigma of fetishism, perversion, or representation as a third gender or sex. Enterprising and persuasive individuals have always strategized to successfully manipulate the standard criteria and protocol in order to obtain particular hormones and surgeries without committing to a transsexual identity. Based on a 1996 article that appeared in IFGE's *Transgender/Tapestry* magazine, transgender, as a gender identity category similar to that espoused by Virginia Prince in the 1970s, was resurrected and presented in Holland as an established alternative to the transsexual and transvestite classifications. The significance of a discrete transgender category in Holland is inexorably linked to issues of legitimation and the pressing, pragmatic concerns of procuring medical intervention and government subsidized services. Hazel, a transplanted, transgender identified former resident of Amsterdam currently residing in Boston, offers a comparison of the transgender scene in both countries:

> The first time I entered the club in Amsterdam on a cross-dresser's night, I was asked, *"Are you a transsexual or a transvestite?"* That was the most important question—that is what they wanted to know. They are considered to be two entirely different groups of people and there isn't an in-between group. And I think this is because from the 1970s, transsexuals were very successful in getting all the services and all their needs met. Transvestites, on the other hand, were denied any form of service, even on a self-pay basis, because it was considered to be fetishistic, compulsive, narcissistic—so there's a whole pathologizing way of thinking about that.

Hazel's observations concerning the "shaping of a transsexual identity" in order to receive services were repeated by several Dutch informants. This process can be either conscious—as in the case of manipulative strategizing to obtain certain desired medical interventions—or unconscious, as my ethnographic evidence indicates a variety of experiential processes that reformulate life history events. Hazel goes on to say that

> when I came to the U.S., I suddenly found a transgender scene that was much more open, and much better connected to the gay scene. There was a much bigger platform where I could be myself. I think the transgender rhetoric in the U.S. is a better way of thinking because it allows people to identify with something that has a very broad meaning.

Hazel alludes to the possibility of a discourse as the defining element of identity. As a transgendered individual in Amsterdam, Hazel experiences herself as disenfranchised, since she identifies with neither the transvestite nor the transsexual identities available to her. Transgender provides a "bigger platform" and a "broad meaning," allowing space for interpretation and exploration.

For Arthur—activist, lobbyist, and perhaps the foremost exponent of transgenderism in Holland today—transgender as an identity category represents a compromise, a "choice that had to be made in order to keep several important parts of my life together." As a solid, middle-class businessman, Arthur is married and the father of two sons. Personally credited with introducing the term "transgender" into Dutch gender discourse two years ago, Arthur had initially identified himself as a transvestite. Through a progressive transformation of identity coinciding with a desire to begin hormone treatment to feminize his body, Arthur views his situation from a pragmatic perspective. Apparently, Arthur had been "evaluated" by the gender team and found to be an appropriate client for sexual reassignment surgery. In other words, he was considered by the experts to meet transsexual criteria. Arthur indicated that should his wife die or leave him and should he find himself in other circumstances, he might not be opposed to surgical intervention and a complete transformation. For the last two years, he has been battling Dutch medical and legislative bodies for the right for transgendered individuals to obtain hormones and psychological services as part of their subsidized routine medical benefits.

Arthur espouses a discourse of compromise where, faced with the reality of life's commitments, he pursues, or in this case creates, an alternative gender ideology. Rather than electing to express a transsexual identity for practical reasons, Arthur discovered a gender classification in *Transgender/Tapestry* magazine—transgenderism—which validates his conceptions of his gendered self and enables him to realize aspects of his life in a different and valuable way. In his medical and legal battle for recognition and acceptance, Arthur is actively formulating a discourse that will provide individuals with an alternative category, thereby enabling them to renegotiate and shape their perceptions and gendered life histories according to this new ideology.

Resistance to transgender interventions has been expressed by some members of the Dutch medical establishment. Primarily, they are concerned with the sociocultural repercussions of introducing change too rapidly in a sex/gender dichotomous society and the consequences of creating what they allude to as a third sex/gender category. How does society classify a genetic male who uses hormones to stimulate breast development while still choosing to retain his penis? Or, despite the individual's own definition of masculinity, what category is used to describe a genetic female who undergoes hormone therapy, mastectomy, and hysterectomy but refuses to relinquish vaginal sensation by undergoing a phalloplasty? As Anne Bolin points out in her essay "Traversing Gender," these "rebellious bodies" or "hybridizations create disarray" as they "threaten to overthrow the *biopower* of the medical orthodoxy." The formation of "she/males" and the diverse permutations of this combination will certainly affect Dutch family law, should an individual apply to have official records, including the birth certificate, changed to meet a new or predominant gender/sex identity. As the law now stands, adherence to a complete surgical and hormonal program is necessary for legal changes in identification. This includes the individual's agreement to compulsory castration or hysterectomy, if necessary, to prevent production or bearing of children.

Recently, Dutch government-sponsored research has focused on attempts to use accepted quantitative and qualitative methodological paradigms to carve out a unique and, in this case, specific identity category of transgenderism. These studies focus on the psychological and social criteria necessitating the formation of a transgender identity and the behaviors that distinguish transgendered individuals from transvestites and classical transsexuals. A goal is to develop a protocol for the selection of candidates whose psychological profile and life circumstances warrant the granting of healthcare reimbursements for hormone therapy and surgical treatment.

In conclusion, category paradigms and their behavioral contents are limited at best since they always reflect an artificial rigidity and delineation that is not realized in human experience. The idea that, for diagnostic and treatment purposes, identities must be perceived as fixed ignores the ambiguity and instability that encapsulate the gender discourses of my informants. Ethnographic evidence strongly suggests that identity is a fluid construction, comprised of dynamic processes and multiple motivations that are dependent on context, situational constraints, and life course events. In my discussions with Dutch researchers, I found that most were completely unaware of the cycle of marginalization and disenfranchisement that the government-sponsored research has the potential to create for other gender alternative people who fail to fit the new model of the classical transgender. Although seemingly benign, the conceptualization of a new gender category creates a discursive identity with the power to validate and legitimate while simultaneously suppressing alternative and idiosyncratic gender ideologies.

References

Bolin, Ann. (1988). *In search of Eve: Transsexual rites of passage,* Westport, CT: Bergin and Garvey.

Prince, Virginia. (1976). *Understanding cross-dressing.* Los Angeles: Chevalier Publications.

The Art and Nature of Gender

Jamison Green

Gender theory today is a triumph of feminism. Without the feminist movement we could not discuss gender because, according to the dominance and oppression model our intellectual culture has developed, the oppressors would be incapable of formulating the concepts necessary to explore the subject. Why should they bother, when their patriarchal system only benefits by maintaining itself?

Thanks to the feminist critique, we can now say "gender is a social construction," as if we are above it all, and we rail against the very creation of gender as a system of oppression. Some transgender activists and gender theorists both (though possibly for very different reasons!) advocate the eradication of gender to free us from the bonds of enforced arbitrary behavioral norms predicated on gender. We assume that gender is "a system of meaning that consists of two opposite and exclusive categories in which all people are placed, [that] gender is based on a cultural interpretation of the biological differences between men and women—not the biological differences themselves" (Kulick, 1987:11). Further, we accept Kulick's premise that biology is always perceived through culture, through translation in terms of symbols and ideas that we use in our daily and ritual life.

While I think the preceding definition reflects the prevailing cultural construct of gender, the truth of our present social reality in Western thought, I also think it is too simple.

We are sensate beings, and we interpret the world through all our senses, not only through our analytical minds in which we apply the labels and value judgments of our time. As soon as we try to speak about our experience it becomes filtered, removed into the conscious realm where it can be manipulated, edited, and revised to suit our egos, build careers, dance around the issues, or flay the competition. We know that language—words—are power, much like gender can be when we use or manipulate it consciously. Perhaps we take our language about gender too seriously.

If we agree that "man" and "woman" are words that signify

male-bodied or female-bodied people (respectively), then we may also agree that the terms "man" and "male" and "woman" and "female" refer to specific types of bodies (or representations of those bodies) in a way that connotes a social role, such as the wearing of certain clothing that indicates the body. But through our agreements we are not creating a capacity to observe the gender of the people to whom we are referring. It is a logical leap to suppose that the man who is male is also masculine, or the woman who is female is also feminine, regardless of the clothing he or she is wearing. If gender distinguishes the "nonbiological features resulting from a person's ascribed status of either female or male" (Doyle and Paludi, 1998:6), then gender studies are focused on the social differences between persons with (presumably) male and female bodies. This is not really talking about gender, but about sociology and politics. Like the notion that "Femininity unfolds naturally, whereas masculinity must be achieved..." (Herdt, 1994), our biases about gender are rooted in biological theory, sociobiological functions, and the extent of our resistance to these biases. When are we going to really talk about gender? Not until we learn to separate gender from the language we have traditionally used to describe it. So, just as we have been working to separate sex and gender (and to avoid using those terms interchangeably), we must next attempt to wrest gender loose from the grip of social constructionism to see whether it still exists and can withstand examination on its own.

We find it easy to trivialize gender expression. The very term "gender role" connotes its pliability, its lack of serious reality for students of social science and others concerned with definitions and social processes. The baseline values of Western culture have encouraged us to trivialize gender diversity as a means of retaining social order, particularly with respect to sexual behavior, as reflected in the widespread assumptions that men with feminine characteristics are homosexual, women with masculine characteristics are lesbian (and all lesbians want to be men!), and men who cross-dress are homosexual (that is, they want to be women). Here the conflation of assumed meanings for bodies and derived signals for gender becomes the basis for misinformation. For someone who does not understand gender diversity there is supposed to be a correlation between a body and its gender, and all that gender is "supposed" to stand for: sexual orientation, sexual behavior, physical appearance, social role.

To describe an instance in which I believe gender does not follow the prescriptive social construction, I offer a personal experience. From 1973 through 1976, I was employed as a construction cable splicer by a regional Bell Telephone System company in the U.S.A. I was the first female-bodied individual to successfully complete the physical training program for this position, a training so rigorous that the majority of prescreened and otherwise qualified men regularly failed to pass. During this period, I also decided to let my hair grow longer in an effort to confound a Bell System regulation against long hair for splicers (all of whom were presumed by the regulations to be male), just to see whether the administration would ask me to cut it. They did not ask this, tacitly creating the opportunity for the men to let their hair grow, too. What surprised me was that the longer my hair grew, the more consistently strangers perceived me as male. I had assumed that by letting my hair grow I would have to accept that I was imposing a more stable female pattern onto the senses of others who would be attributing gender to me. I came to understand that gender cues are not all visual, superficial, or visible on the body. They are not all determined by cultural presumptions or agreements about the variable characteristics that we take on as roles or costumes. I had a growing sense that there is something about gender that is not expressed only in clothing, hairstyles, body shapes, voices, or even the conscious awareness that a body has a designated sex. There is something else going on.

Around this time (in the mid-1970s) my friends and I became aware that female-to-male transsexualism was possible. Among most lesbians I knew at the time, the concept of "women becoming men" was abhorrent, and many women began holding me up as an example of the kind of strong woman that transsexuals could be if they would only let themselves. A physician friend discussed with me a strategy of offering myself to a gender clinic to see whether they would diagnose me as a transsexual because of my masculine character and demeanor. Of course, I could not admit to her, a fellow lesbian, that I actually entertained the notion of surgical sex reassignment. Her fantasy was that once diagnosed I would then stand up and shout "Ah-*ha*! You fools!! Do you think just because I am perceived as a male and hold a 'man's' job that I will automatically want to be classified physically and socially as a male? What is wrong with being a woman with masculine characteristics?" As soon as we had constructed this scenario, I began to question whether this was the proper statement for me to make. I was comfortable with the label of "lesbian," but I was decidedly uncomfortable with the label of "woman," and I confided to my friend that I didn't think I was fit for the assignment of exposing the gender clinics because I feared that if offered the opportunity to change my body I would actually appreciate doing so. I sheepishly confessed that in fact I perceived myself as more male than female, knowing I was speaking heresy.

"What do you mean?" my friend inquired.

"I mean I would like to have a male body," I replied.

"Well, so would I, in some ways," she said. "But you wouldn't want to really be a man, would you?"

"What do you mean?" I asked.

"I mean, it is so much better to be a woman, don't you think?"

"But I don't think I am a woman," I said. "I don't think I was, or ever could or will be." And I realized in that moment that I wanted to be *some*thing, to be physically in the world unquestionably, both for myself and for others. All the theorizing about gender, about male privilege or the moral superiority of women, about the appropriateness of altering one's body, or the false dichotomy that the dialectic of apparent biological sex and social gender force onto our bodies and souls would never neutralize that need in me. I also knew that whether or not I did change my body, I would always be not male and not female, unlike other people who never disturbed the social waters around them or who never thought to question where they fitted in. I would always be different from conventionally gendered beings. And ultimately, by changing my appearance to reflect my masculine gender identity, I did not narrow my perspective to obliterate the feminine, but in fact I broadened my own understanding of what it can mean socially to be labeled man or woman. And since then I have dedicated myself to creating a social space for men with female histories, and for men with or without female histories who do not buy into a traditional gender role, regardless of their appearance.

Appearance has a lot to do with how we perceive gender and the kind of attributes we assign to people upon first meeting them. Kessler and McKenna provide a thoughtful and comprehensive analysis of this process, noting that "biology provides 'signs' for us. Signs are not gender, but they serve as 'good reasons' for our attributions in a world where biological facts are seen as the ultimate reasons" (1978:4). In this framework, it is reasonable to assume that I am a man and that I enjoy all the customary attributes of the masculine gender within the culture in which I live. This is an "incorrigible proposition" (Kessler and McKenna, 1978:43)[1] defining the interpretive structure of the onlooker, the one who is analyzing. My own experience is somewhat different from what might "reasonably" be assumed, and it is here, in the realm of the Subject, that I propose that gender resides. I believe gender belongs to each individual, to do with as he or she pleases: it is not possible for an "objective" observer to paste gender on another person by labeling them with a gender that the person does not feel, whether or not that gender is expressed. Any attempt to make another person's gender an object that can be isolated and analyzed is doomed to failure, not because gender is a social construction and so requires more than one person to create it, but because gender is a private matter that may be publicly expressed and it cannot be abstracted from the Subject's (conscious or unconscious) control.

If gender were a social construction, it would require, like language, a "speaker" and a "listener." It is between these two actors that gender would be defined, negotiated, corroborated, or challenged, and without this interaction there would be no need for gender. This is like saying that if a tree fell in the forest and no one was there to hear the sound, then there was no sound, or perhaps no tree actually fell! One "incorrigible proposition" concerning gender today is that it depends on biology; another "incorrigible proposition" is that it depends on someone else's interpretation. We all have the ability to speak and the ability to express and interpret gender, regardless of our apparent biology or whether or not anyone is there to hear us.

Analyzing the distribution of power between men and women in Western culture, it is perfectly reasonable to see how women could appeal to gender. Such women were arguing in the 1960s and 1970s (and earlier) that women's lives should not be circumscribed by their ability to bear children, and women should not be forced to take on other socially defined tasks specifically because their female body was assumed to predispose them to those tasks. As women distanced themselves from their biological "roles," gender seemed to evolve into the only differentiation factor that was safe to talk about. When menstruation was viewed as an incorrigible proposition supporting the belief that women were incapable of being responsible voters or holding responsible positions in the workforce, what better time to haul out the gender ammunition, forcing the discussion away from the physicality of the body? This is a viable political strategy that shifts the debate and widens the field for potential allies. We used the same technique when we instituted "transgender" as an overarching term, minimizing the stranglehold of the medical/psychiatric profession (and the law's reliance on that profession) on transsexual lives and broadening the scope of gender variance to include virtually anyone. So why does contemporary gender theory still cling to the traditional gender-power rhetoric? Why are transpeople being held responsible for preserving the dichotomy between the genders? Because of transsexualism's perceived threat to the presumed reliability—or naturalness—of the body.

Kessler and McKenna (1978:20) stated that "Science will soon be able to construct perfectly functioning penises. Because of this we will never know what would have been the long-range repercussions, on concepts about gender, of having a group of men in society who do not have penises." That was twenty years ago, and those "perfectly functioning penises" have yet to be surgically attached to female to male (FtM) bodies. Even if they were actually available, I question the logic that dictates all FtMs would want them. Many transsexual men feel that they *do* have perfectly functioning penises without surgery, or with modest surgical modification of the native organs. Some of their partners might not be so interested in perfectly functioning penises, either. The presence of a penis may bestow male status on an infant, but these adult men know that it is not a penis that makes them male. Here is a group of men who do not achieve gender membership through their genitals, yet these very men are still accused of reinforcing the gender status quo simply because they are transsexual and appear masculine in other ways, such as beard growth and musculature.

What is so wrong with being masculine? No one asks transsexual men about their politics; very few of them are asked about the ways in which they manifest sexual desire. How would anyone know whether these men are "buying into stereotypical gender roles," or whether they simply fit some categorical gender stereotypes and don't fit others, just like anyone else? It is apparently much more satisfying for researchers to take small samples and ask leading questions, to seek out statements that use "typical" transsexual validation language, such as, "I always felt more like a boy than a girl," and to interpret these statements as a conscious effort to create a sense of the naturalness of the transsexual subject's gender, the assumption being that gender is not natural and the body is natural, and an altered body is no longer natural, while an unaltered gender is impossible since gender is socially constructed. This belief structure is used to justify viewing transsexual people as a class of individuals who are incapable of assailing the paradigmatic ramparts. Never mind that transsexual people often learn that certain language is more effective in eliciting sympathy for their condition, thus ensuring their physical and psychic survival in a world that is resolutely hostile to variation or surprise. It is comparable to saying that because a woman has a woman's body she must be a woman and conform to the gender role assigned to her by others. It is comparable to saying biology is not destiny, unless you are transsexual, or unless you have a penis, or maybe unless you don't have a penis. It's all right for a woman to be masculine, or a man to be feminine—to a point! But just try to take on some other gendered embodiment to find out how dismissive people can be, or how rigid are their concepts of gender, of identity, or of the immutable landscape of the body.

Non-transsexual interpretations of transsexual and transgender expression, both in physical space and in language, often reflect an easy dismissal of transpersons' agency, if not their very existence (Hale, 1998:108–113). In the rush to eliminate transsexuals from the social landscape, theorists (and sometimes former friends) reify the sex/gender paradigm they purport to resist. It is true that I have chosen the option of transsexual treatment in order to obtain the social and legal status of male because that is the place in contemporary society in which I feel most at home, most comfortable, most natural, and most creative. I felt that transsexual treatment was the only means by which I could achieve my personal goal of adulthood. This means that I have chosen to change my appearance, something many people do in many ways. It does not mean that my gender is socially—or even medically—constructed. My gender has not changed; I have simply made its message clear.

Non-transsexual theorists do not always recognize the fact that when transsexual people undertake this type of physical change, we do not know what form our new physical manifestation will take, that we cannot will ourselves to look like perfect incarnations of masculinity or femininity. Just like non-transsexual people we must accept ourselves the way we are (to paraphrase the classic prayer), changing the things we can, accepting the things we can't change, and acquiring the wisdom to know the difference. And those of us who do grow up into adult men or women—referring now *only* to appearance—do not need to bear any greater share of the social burden for the dichotomy between the two dominant forms of gender expression than any other person who appears normatively female or male. The theory that if transsexual people had some other culturally constructed option, a place to be socially male (or female, or whatever!) while remaining physically female (or male), then we would categorically refuse body-altering technology is pure utopian conjecture. Such a conjecture is based upon some particular incorrigible propositions, or perhaps (for those who already appear normatively gendered) rooted in the desire to step outside the limitations of socially constructed gender without changing one's appearance. This is another way of proving that the people who appear different or non-normative (read transgendered) are deluded for wanting to appear "normal," while the "normal" people can feel at least normal, or possibly even superior (if only politically correct), for wanting (or being willing) to appear different.

Transsexualism, for non-transsexual gender theorists, somehow seems to represent a paradoxical struggle with what is natural and what is artificial, and the values invested in either category. In Western culture, there is a belief, drawn, I think, from humankind's unwavering fascination with itself, that Art is always an effort to understand or control Nature. In this model, Nature is viewed as the world separate from human consciousness and capability. Both the forces of Nature and the human capacity for creativity were made manifest by divine creation, with "man" in the image of God and therefore the highest creation. Thus, the systems or objects that human beings create become an attempt to place comprehensible controls on an arbitrary and capricious natural world and become invested with a higher value than the "native" world. Today, however, we are also informed by the neo-science of ecology and by a new appreciation of non-Western cultures, which may have had slightly different gender categories, though the dichotomy of male and female has existed in virtually every culture. Today we ascribe a high value to human activities that run in concert with rather than dominate Nature.

The two ideologies represented by these old and new worldviews become confused when transsexualism is brought into the mix because of assumptions about transsexual agency, gender/power struggles, the primacy of the visible body, and our very incomplete knowledge about biological sex differentiation. We forget about our private, individual efforts to be more than our bodies; and we forget that we are part of the natural world. We forget that Art may be a way to more than order, that Art can help us to open ourselves to the emotional power of Nature. We become blind to the incredible diversity that exists in the natural world, and we still so easily employ a divide-and-conquer strategy to organize our ideas and separate ourselves from the ideologies with which, and people with whom, we don't feel comfortable.

We are part of the natural world, and our artfulness permits us to change that world. When it comes to gender, if we are invested in the dichotomy between masculine and feminine as a system of socially constructed power distribution, we will never be able to value the naturalness or artfulness of an individual's gender expression because we will always suspect the individual's motives: anything out of the ordinary must be a quest for power, and is therefore victimizing someone. But gender, like race, is not a power system in itself; gender, like race or like language, is a physical trait that some people use to gain or distribute power. Like language, our gender is both natural and artificial; the ability to have gender and language both reside in the natural or native beingness of individuals, whether their expressed gender reinforces, contradicts, or is randomly confused by an observer's cultural concepts of their bodies, or whether their ability to speak is compromised by physical deformity, or they happen to speak a different language than their listener can comprehend.

Art begins when we begin to exercise power—not power over others, but the power of creativity, the power to increase vocabulary, to change our environment, our clothing, or our bodies, to learn new dialects or new tongues that give us versatility and greater mobility, allowing us to travel across boundaries and to broaden our perspective. In the traditional paradigm of dichotomies, movement like this means being forced to swear allegiance to the new state of residence or forever being viewed as an outsider if one's identity papers were not issued by the local authorities. We are ignoring the realities of contemporary life: fused art forms and mixed media challenge and stimulate our perceptions; jet travel and the internet take us farther than ever before; there are lesbians falling in love with FtMs, and lesbians falling in love with MtFs, and with non-transsexual men; there are women with penises, men with vaginas, and the reality of people with genitalia that are virtually intersexed.

Our world is not made up of easy dichotomies, of simple hardware and software. At the interface of the programming and the machine, there is an operating system that is composed of a special type of language, and as a culture we don't understand this interface very well. Some programmers and engineers understand it as this analogy applies to computers. But in the world of human Art and Nature, mind and body, we do not understand this interface very well at all. We call what we don't understand at this particular border "soul," "psyche," or various "disorders," and we endeavor to structure our observations of these phenomena to fit our incorrigible propositions. This is how gender became the sign or signal of sex or sexual orientation, and how gender politics and gender power became the battlefield on which we wage the war of equality between the sexes: When gender became the focal point for discussion about social roles, the intellectual culture theorists lost sight of what gender is for an individual and how the centrality of gender identity informs each of us independently, without a need for social validation.

In looking at children, as Kessler and McKenna (1978, ch. 4) show in their overview of several major studies, psychologists set up interview situations or establish clinical studies in which they ask for a child's interpretations of bodies. What they seem to be looking for are the unadulterated social cues that signify man and woman. What they find is that children often see things differently than adults do, or at least have language that expresses things differently than adults would, but that children all too easily can be made to comprehend adult language about gender and adapt to adult constructions of gender as they adapt to most adult dictates. I found it interesting that Kessler and McKenna (1978:167) note that several transsexuals have mentioned to us that they have the most difficulty "passing" with young children. These are all MtF transsexual people who presumably retain some masculine physical traits that are confusing to children who have learned what is male and female in appearance by generalizing from their environments. This point seems intended to cast aspersions on the "reality" of the transsexual peoples' gender identity, especially since the doubt seems to come from the transsexual subjects themselves. My experience with children as a female-bodied transgendered adult was that children most frequently ascribed masculine gender to me, or, if uncertain, they would ask whether I was a boy or a girl. I noticed that if I told them I was a girl, they would doubt me. If I told them I was a boy, they felt reassured, but then some of them would continue to question me, prompted by some physical characteristic such as the presence of breasts or the lack of a beard. Since I have physically changed to a more obviously male presentation, no children have been troubled enough to ask about my sex. I think it is a faulty presumption to think that children have any special knowledge or awareness when it comes to gender or sex: children are very busy trying to acquire knowledge of the relationships between themselves and their surroundings, and often their very survival depends on pleasing the adults around them. As human beings, children are adept at categorizing, and what seems to be gender awareness in children could easily also be the gross perception of normative versus non-normative visual cues, which does not inherently invalidate those non-normative signs at all.

It is reasonable to conjecture that the organizing principles that drive children's gender-segregating behaviors (at least before the age of five) may not depend on an individual child's self-ascribed gender identity (which he or she is most probably incapable of expressing in words). More likely, this drive depends entirely upon adult expectations that are dictated by genitalia and the conflation of sex and gender as systems of meaning that depend on and reinforce one another. Again, so long as we think of gender only as an arbitrary and culturally variable system of role behavior designed to reinforce biological sex differences and the distribution of social power, it is easy to dismiss gender difference and any transgendered individual's personal struggle to come to terms with that difference.

To begin to formulate a new paradigm, what if we called gender the Art of self-expression? What if gender studies were not the analysis of differences between male-bodied and female-bodied individuals as interpreted through culture and reflected in patriarchal dominance, but instead became the analysis of the many and various ways we can and do express ourselves as individual human beings, and of how our assumptions serve or do not serve us in social or psychological ways? What if gender was a nonverbal language, a way in which we tell others something about ourselves before we engage in verbal or physical interaction or intimacy?

If gender is a type of language, some very adept individuals are capable of speaking many dialects, as well as many different derivative languages. Cross-dressers, drag kings and queens, androgynous transgendered people, butches and femmes of either sex, and many transsexual people may all draw upon gender interpretation skills that non-transgendered people don't have, or have not cultivated, or find unnatural. Granted, some transsexual people are very rigid about their own gender identification, but that may be because they have had to fight very hard to assert it, or because they've learned a certain kind of behavior is expected of them. For many people—not only transsexual people—gender expression is an important survival mechanism. When we express judgment about a person's gender expression, whether that judgment comes from our own conservatism (supporting a rigid gender dichotomy) or liberalism (supporting a wide variety of gender expression), we are effectively expressing a lack of tolerance for diversity and appreciation of individuality.

What if gender was the interface between our psyche and our cognitive mind/body/sex? What if gender was an aspect of personality, rather than an aspect of sexuality, an aspect of sexual desire rather than reproductive function? What if it was not gender that gave certain individuals certain types of power in society, but actual bodies in concert with other arbitrary factors like heritage, wealth, special skills or abilities? By focusing on the proposition that dichotomous gender is the bellwether of social privilege, and by viewing transsexual people as social constructions who learn or manipulate gender conventions, we deny the incredible potential of gender variance and its natural diversity, and we categorically deny individuals' agency in experiencing or freely expressing their own genders.

Both the Art and the Nature of gender reside in the individual. The current social constructions of gender are based on interpretations of the body and the social roles assigned to bodies. If an individual's sense of his or her own gender relied exclusively on how others read his or her body and reacted to it, then it would likely follow as true that if you assigned an infant's sex, fixed its genitals to match that sex, and raised the child in the manner of other bodies of that type, the child should have no problem adopting that social role.

This scenario does not work, as the John/Joan case has illustrated.[2] I contend that gender cannot be assigned by others, but only interpreted—or misinterpreted.

Under the current paradigm, gender is a red herring distracting us from the effort to shatter the hegemony of the patriarchy. Gender is not sex, not the body, not a social role. It is, to me, an energy that is pre-sexual, that informs but does not dictate sexuality, that communicates something about the psyche, not the soma.

To understand gender, we must stop trying to equate gender with genitals. We must defuse our concept of gender in order to obliterate the power of imposed roles and schemata that don't reflect experience—for both transsexual and non-transsexual people. In attempting to eradicate gender as a strategy for eliminating the hierarchies inherent in the current social constructions of sex and bodies, we are enforcing a harsh censorship, imposing a terrible silence on an important part of human individuality.

References

Colapinto, John. (1997). The true story of John/Joan. *Rolling Stone Magazine*, no. 775, pp. 54–97.

Doyle, James A. and Michele A. Paludi. (1998). *Sex and gender: The human experience*, 4th edition. New York: McGraw-Hill.

Hale, C. Jacob. (1998). Tracing a ghostly memory in my throat. In *Men doing feminism*, Tom Digby (ed.). New York and London: Routledge, 108–113.

Herdt, Gilbert (ed.). (1994). *Third sex, third gender: Beyond sexual dimorphism in culture and history*. New York: Zone Books.

Kessler, Suzanne J. (1998). *Lessons from the intersexed*. Piscataway, NJ, and London: Rutgers University Press.

Kessler, Suzanne J. and Wendy McKenna. (1978). *Gender: An ethnomethodological approach*. Chicago: University of Chicago Press.

Kulick, Don (ed.). (1987). *Fran kon till genus (From sex to gender)*. Boras, Sweden: Carlssons.

Notes

©1998 by Jamison Green. A summary of this paper was presented at the Third International Congress on Sex and Gender, 19 September 1998, Exeter College, Oxford University, England.

1. "Incorrigible propositions" are the beliefs we hold as defining what is real.
2. See the recent work of Dr. Milton Diamond, a popular rendition of which appeared in *Rolling Stone Magazine*, No. 775. See also Kessler, 1998:6.

Enactments of Difference

Tarquam McKenna

Gay men and lesbian women have been a hidden minority within academe and the schooling community. In some instances gay people have been tolerated and twittered about as their sexual orientation has been queried. These men and women have been forced into silences about their true selves, adhering to the hegemonic way of being and practicing heteronormative ways of encountering the world. Encounters in academe have not always been profitable for gay/lesbian people. The worlds of gay/lesbian educators and students alike are still not considered as an arena of respectable academic inquiry. These people are outside the inner ring of "inclusive education." However in the last five years the trend has begun to shift toward hearing their narratives. The outsiders are entering the world of academic inquiry. This causes agitation in the minds of many readers but I hold this agitation is a path that the androgynous community will also need to tread when we are contemplating the androgynous person in education.

There are two main themes in this chapter—performativity, or "doing gay" and the essential truths of gay/lesbian educators, or "being gay" with attention to "how this happens in schools." The ideas here are necessarily incomplete as the boundaries of these gay/lesbian people are constantly shifting and have shifted since I presented this chapter as a paper in 1998.[1] Alongside some public narratives from elsewhere, I present conversations with Australian teachers whom I shall call Jay, Alexis, Barbara, Alain and Damian.

The marginalization of straight, gay, lesbian, queer, transgender, transsexual, bisexual (and all other gradations of sexuality and gender of) people has led to a sense of self-loathing and hatred that is documented elsewhere and known as internalized homophobia.

> That sense of self loathing is present every day. I went into work as a "first-year-out" school teacher working in a country school—where it was all about sport as a social nucleus on the weekends and during the evenings. I felt the insecurities of the novice, as well I felt I couldn't join into the work

community because I was unable to be something, be enough—be accepted. Of course this was only part of the picture, part of the isolation that I was experiencing. I hated the word "gay"—it was something I was labeled as I was growing up and the idea of being a "gay" teacher was really the impossible—it would have further fractured my sense of being, my day-to-day functioning. (Jay)

We were shunned by many of our classmates for being, as many saw us, just plain weird...We were also picked on. We were called queer and faggot and a host of other homophobic slurs. We were also used as punching bags by our classmates just for being different, something that sent us into further isolation. (Alexis)

The breaking of this sense of being viewed as morally soiled, an outcast or a political threat is the tale of this chapter. I support the inclusion of gay/lesbian people in this book as those who are androgynous might learn through sharing their conversational ideals and their experiences of "repulsions."[2] Inclusion should not have to be "purchased" at such great costs. The inclusion of gay/lesbian people in this book is based on a fundamental right to dignity: dignity that should be afforded to all people. Beyond the values that these people represent which are most strongly rooted in heterosexual minds by their "corporeal activities" there is the need to recognize that the most powerful argument for the inclusion of gay/lesbian people in all aspects of existence is that their right to exist is a moral one.

That internalized homophobia to me as a teacher means I still think about the notion of the pedophile having access to the innocents—their vulnerability or my apparent power in this area. It is quite irrational but there is still such a mythic dimension around this issue. My own experience is that I was molested as a child so it sits in a very potent space for me and this figures in my internalizing it as an issue. (Jay)

I'm 17 and I'm gay. Adolescence is hell for me. I am told that my sexuality is something to be ashamed of, something to hide, something evil. I have cowered in my closet in shame and fear. I found myself lying to my parents and friends, being constantly afraid of discovery, and censoring my words and actions with paranoid concentration. I remember hiding books from my parents because I was ashamed of them discovering about me. In short, I hated my sexuality and myself. My closet wasn't a refuge, it was a prison, and it was destroying me. By staying silent, I was confirming the emotions that were killing me inside. I am not just a statistic. I live in a Boston suburb in a white house with black shutters. I go to school every day, feeling that I can't be honest, that I have no right to be proud, that I am a second-class citizen. Just this past week, as I was walking down my street in my town where I have lived all of my life, a pick-up truck full of guys ran me off the road, screaming "You lesbian!" at me. Homophobia is everywhere, and bigotry is inexcusable. It's time to start showing you care.[3]

I have especially wanted to question politics, to bring to light in the political field, as in the historical and philosophical field interrogation, some problem that had not been recognized before. (Foucault, 1984:374)

Foucault invites questions and calls for "bringing into the light" the darkness of not being recognized. Damian elaborates on this hiding in the shadow realms:

I have had enough of hiding. I guess it is all right not to be out but the cost is too much when I go home. I'm 33 now and I should be able to be comfortable, shouldn't I? I get quite scared when my mum asks me about my life and says things like when will I "get married". Perhaps the risk of coming out is still too big for me. She doesn't see that Neil and I have been together for the last 7 years does she? That's not her idea of marriage. (Damian)

Not all recognition is easily accommodated, as being seen brings anxiety and fear for the hitherto alienated or hidden gay/lesbian people. There is disquiet and unease in being visible.

At Gay Pride I was terrified as there were so many people. I was taken up with the excitement and scared at the same time that someone would see me. I was really frightened at the point when you grabbed my hand and just took it and we walked behind the floats. My heart was beating hard. (Barbara)

Foucault suggests that a "problem" can be brought into awareness through recognition. Those thus recognized do, however, suggest that conditioning and self-doubt make such awareness challenging. This is illustrated in Alain's narrative:

I don't know what I would call myself. If you think I am gay then that is all right but I don't like to be recognized as just gay. It is probably that I am embarrassed or ashamed that I cannot bear the idea of being recognized. I wonder if it would be possible to be recognized as gay and that is good? (Alain)

In San Francisco there are ghettos of gay people. The Castro area. Is that what you mean when you talk about our world or do you mean something else? The word is hard to understand as our world is not different and different at the same time. I don't like cloning and just going to gay places. That happens in Sydney or New York but here we don't have that much "gay only" places. (Damian)

The language of hate that applies offensive social labels to gays[4] lesbians[5] and transgendered people[6] has been married to acts of homophobic oppression that the oppressed now see as "inevitable." Perhaps it is fortunate that intersex people are so invisible that they have not yet attracted a similar specialized vocabulary.

> I have been called lots of things and once got bashed up in Alexander Library Square. I was walking alone and that was silly but I thought it isn't too late. They mugged me, calling me names, and they bashed me...not taking any money. (Alexis)

Socially sanctioned violence is still anticipated by gay/lesbian people, who are victims of senseless hate crimes. Heteronormative practices include assault, vilification, and discrimination. This problem has been recognized before and still demands solution. The agents that will bring answers to the questions of political inclusion are those gay/lesbian people who themselves know their oppression—both internalized and externally derived. These people deserve to be brought into the light so their worlds can be heard, seen, and brought to a place where they feel a sense of their sanity,[7] safety,[8] and confidence[9] in being who they are.

The gay/lesbian people speaking throughout this chapter and their cultures are part of *our* world and we need to listen to their tales about how we have construed their selfhood. These conversations need to be heard for the act of speaking is breaking the silences of being unheard. The emphasis of this chapter is less on normalizing practices than on those practices which deny selfhood to those who are gay/lesbian. These people identify through their narratives those practices that they have perceived alienating.

> I am so careful about how I express my personal life, actually any life lived outside the confines of work—there is such a pit of feeling alienated. I can't talk about my partner, or casually speak about what I have been reading if it is of a "queer" nature (even that term doesn't sit well with me), what I did over the weekend, etcetera. I once dated a colleague of mine and it began as a nervous adventure, a secret game, and it developed into a more grounded friendship, but what I was so aware of was his own sense of alienation, his sense of threat and danger from our friendship/relationship—it feels on this level a "soiling" of something which was quite lovely. (Jay)

The author is present as a receiver of silenced voices. This chapter considers how silence about homosexuality and the marking of gay/lesbian sexuality flow through these people's lives. One task is to describe and define the silences, the ignorances, and the notions of difference that continue to naturalize the male/female divide for them. The more poignant confessional task is in hearing what maladaptive action has been inflicted on them as they become the people they are. In effect, excluding gays/lesbians from conversations and their rites of identity and failing to treat them with respect appear to me to constitute a violation of the justice that should be afforded to all people.

> I have a senior colleague who is obvious about his gay sexuality; he is more flamboyant and flirts with the idea that people will find him amusing (in a Noel Coward sort of way, I guess). I really feel that if I was more obvious about my sexuality it would be easier for other people to deal with me—it can be a way of containing their own fears or prejudices, in that they can treat me in a certain way that they are used to. In this way there is a sense of dislocation—that is not too strong a word or feeling—I feel torn by this ambiguous sense of being neither "here" nor "there". There is a paradigm of gayness that is so clearly communicated out there in our culture that I am really not too comfortable with. In gay culture there is such a fashion driven sense of what it means to be gay that it takes sometimes a huge leap into "being gay" that you can feel the outsider there as well. I'm talking about things like age, nationality, physical build, dick size, or clothing. (Jay)

The gay/lesbian people in this chapter become dislocated as the culture they are part of is founded on ideologies and theories of binary opposition—they are always seen as "outsiders" who serve to present themselves to the prevailing heteronormative viewpoint. Their worlds and ways of being generate agitations or currents that have declared themselves as "movements" which only now[10] are forcefully beginning to redress the rigidities and inequalities arising from the binaries of the sex/gender system; especially the feminine/masculine and homo/hetero-divides.

> It is through articulation...we engage in the concrete in order to change...articulation is not just a process of creating connections, much in the same way that hegemony is not domination but the process of creating and maintaining consensus of co-ordinating interests. (Slack 1996:114).

A key splintered articulation is "false self-behavior [that] involves the speaker in not saying what one thinks or believes," (Slack 1996:117). In their tales of enactment and difference, these gay/lesbian people tell *us who we are*. Their articulations are snippets of narratives and occasions for us to contemplate and enact the rearticulation of who we are. The hazard of exposure, vulnerability, and nakedness that all of us experience as we hear these truths is a step toward our own empowerment: an empowerment that cannot occur until we are moved, shaken, or agitated at our core by the manner in which gay/lesbian people have been maltreated.

The people that we are demonstrate in diverse ways privilege and susceptibility to intense self-consciousness as we interact with varied oppressions and oppressive systems. If we are straight, our sexual identity is hardly open to question. Compulsory heterosexuality (Rich, in Snitow) is built upon the assumption that a standard exists, and all divergence from that standard is impaired behavior.

These conversations with gays/lesbians also reflect other people's intolerance, often leading gay/lesbian people to impulses of self-censorship and guilt, which invariably lead to psychic numbness. All manner of agencies have made less than well-intentioned attempts to "de-privilege" gay/lesbian people but we must also know when and how we do it to ourselves.

> Culture and religion have spun such a big web in relation to my feeling at home with my sexuality—the Catholic notion of sin and shame has featured big in my life. I find that guilt has transferred into my work as a teacher—I am guilty before the judgment of the "school system." I have felt the invisible guilt of the closeted teacher, hearing anecdotes from other colleagues speaking about students they perceive as gay, their experiences of gay people and such. (Jay)

Terry shows how the world of school showed him a way of being in his truth. He says, when he is finally able to talk at the age of 23:

> I went to a school where we wondered about the man who was a little feminine. We all said that he might be homosexual. I wish that I had known a real man who was homosexual as this might have made it easier for me. (Terry)

We will never know if his "role model" was gay or not: though Terry is hoping he was. The enactment of the "man at school" is about telling the young Terry what he might look and act like as a gay man. Terry has used this man as a way to define himself. However, it must be asked why Terry was restricted to such a limited psychic mirror. Jay uses school experiences to define himself as different.

> The school environment imposes certain images which I can never live up to. As a student you had to be good at sport...I wasn't good at sport. So as a teacher I felt very uneasy about that. All the things that went with being a gay student meant that you had to be macho, you had to condemn other people if they were effeminate, and you had to be not gay. That's what it meant to be good at sport. (Jay)

All articulations such as these are of necessity splintered, unstable perceptions, incomplete, understandings, and yet they are all we have to enable mutual understanding of the story. Text is supplemented by daily behavior and a mode of artistic performance and endeavor.

Terry finds that drama and theater can help him approach, uncover and cherish his authentic[11] self. Drama is a way of moving beyond mere textual analysis:

> It was in drama class that I knew I could be whatever I wanted. I developed the ambition to be myself. Ambition is what has got to be there for us to be who we are. I love Madonna as she is so androgynous and I watched a lot of TV as I discovered who I am. She was my role model. Nowadays TV seems to be the thing that tells us who we are most of all. (Terry)

Drama provides an additional field to help reconceptualize this research. As rituals, the arts can hold us to a defined, confined, or refined place of being through *enactment* or *performance* (Bornstein, 1994; Butler, 1993, 1997). In this chapter I return to Peter McLaren's rendering, as he considers the embodiment of being and performance as ways of being. He implies that the act of doing is as valuable as the act of speaking in in-forming identity. Terry is also a transvestite and drag artist who finds meaning in the visual and kinesthetic field of performance art. He uses another name, Candice, in performance. Judith Butler describes performance as the regulations that prescribe "homosexual conduct" where performance and performer are engaged in a "circularity of fabrication and censorship" (1997:107). Candice aspires to use theater and performance as a way to identify with herself as "who she really is."

We are unceasingly subjected to self-reinvention. This is a time of coalescence and fusions of meanings. The reinterpretation of the self occurs through the opportunity afforded Terry to perform in Perth in 1998. Terry as Candice can inscribe her identity and she conforms to a basic need for respect, love, and almost adulation. Terry has determined s/he *can* be fluid:

> I am given a lot of attention when I perform and it helps me to be stronger in myself. People respect me when I am dressed as a woman and they tell me I am pretty. I love the performing part of my life and it is important. The dressing up and getting ready is as important and then I do it in front of all those people. It is so valuable for me to show myself in the shows I do twice a week. (Terry)

Performance as a drag artist recasts the issue of being pretty and shows that particular changes occur in Terry's self-understanding, and disposition. Candice could be saying of herself what Lacan said of truth—"I am the enigma of her who vanishes as soon as she appears" (Lacan, 1977:121).

Candice's audiences are most often gay men. What happens for them? Candice's performance provides other individuals a closer knowing of her: but again she is viewed by the Other as an outsider. The gay audience has "permission to stare" (Goffman) and become almost privileged as theater brings them insight. The art of Candice's performance is the fabrication stating "I am Candice." Gay men can sit before her and the performance becomes an occasion of "self-ascription" (Butler, 1997:107). Their authority is in a fluid place: self-ascribing the "paranoid construal of homosexuality" can make way for more promising self-referral. The performance provides an interrogative encounter for Candice and her audiences alike.

Placing our collective narrative within a current time frame, we can all tell "tales of the field." Through this chapter we can examine the centrality of our multiple truths within a heuristic that requires us to recognize our sexual subjectivities. Terry tells of the re-invention: " I don't think of myself as different...I am just happy to be able to be androgynous and have both Candice and Terry in my life."

I hold that our bodies are a vehicle for the stories that must be spoken, enacted, and drawn as we come to understand our various sexualized selves and worlds: we need to view sexualities as a fluid collection of meanings; sexualities can and must now be expressed by a commitment to cooperative storymaking and to multivoiced accounts that are constructed through relationships we have with ourselves, others, and our world agencies. The act of making the self requires not only the linguistic turn but enactment in drama and other arts. Jay speaks of the value of making the self in visual arts:

> I'm not "out" with my sexuality. This serves me fine as art is a more tacit realm of expression where I can express and articulate these notions. I have the capacity to self-censor and articulate myself concurrently. It is also my approach in teaching art (or design for that matter) where all these "invisible" strands of what we know as our sexuality come into play. The process of making art involves an inner focus and a contemplative mode that takes into account our humanness, which includes our sexual self. (Jay)

Alexis and Damian are two gay men—the former is 33, the latter 38. Alexis's conversation shows his negative self-ascription: "I went to an all boys' school and they were always assuming we were all poofters...it was one of the hardest things to be thought of a poofter."

Being thought of as gay is one of the worst things that can occur in the minds of Australian young adult/adolescent males. Another moment of uncovering *this* difference will be examined more fully later in this chapter. In my work as a teacher trainer with drama students, I hear how frequently "homophobia" or "gays" are talked about from a negative focus; indicating that alienation of gays and other-gendered people is still happening. Damian's self ascription shows the cultural shifts that can occur:

> I was usually teased and frightened to get on the bus. I was called "Romeo" and now I think they'd use another word to tease me. Now Leonardo Di Caprio is Romeo because of the film. I was taunted mercilessly and it was

not making me too happy with myself. (Damian)

Damian recognizes that the frame of the cinema created by the director Baz Luhrmann's *Romeo and Juliet* (1996) can be used to foreground his bodily and psychic shifts in understanding. Damian implies that a historical shift has occurred as Romeo is now seen through the gaze of "spunky" young Leonardo Di Caprio. In Damian's youth, "Romeo" was used by name callers to belittle his effete and effeminate manner. Nowadays cinema, drama, and other performance arts can be a way of glaring at heterocentric beliefs and allow movement toward the denunciation of homophobia. Damian can now "become" Romeo with more cultural comfort as the Romeo of the present day has a very different range of signification from the Romeo of Damian's youth. This performance is for Damian a "renewable action without clear origins...marked by its capacity to break with contents" (Butler, 1997:40). Both Romeo, Damian and Candice have been used in the performance to tread on the borders of the "unsayable." Candice and Damian have a call to explore the boundaries of their legitimacy. For them, performance provides a context for self-definition. Another reflection on a filmic moment illustrates queer fear:

> Teacher: do you recall the film in which the two boys walk into the sunset?
> Student 1 (Female): They weren't holding hands were they, miss?
> Teacher: (Wondering and mistakenly) I think so.
> Student 2 (Male): That'd make em poofters.
> Student 1 (Female): No, they weren't holding hands so they're not gay.
> (A year-10 drama class discussing difference)

Here prejudiced perceptions mimic the stereotypic practices of what they call "poofters"—the imaginary world of the year-10 male student indicates that he has a conditioned perception of difference, which he fears seeing dramatized. It matters little whether the two boys *are* holding hands. This year-10 male does not need or want to see his or others' sexual crises embodied. Even in holding hands. In another cultural context, as Alexis indicates, this would be anything other than inappropriate:

> In Greece all men are that close. Holding hands is seen as OK. I find it hard to believe that Australian men don't understand that but I know when and where to hold hands...like when to kiss or not to. (Alexis)

The kiss and the hand-holding gestures mentioned in these conversations are staged to tell of connection. Alexis is able to recognize where and when they are OK. In discussion of hand-holding the teacher is unwittingly reminding the student that he is fearful of seeing gay behavior *enacted*. This presumes that the two men *are* gay. The 16-year-old male student has volatile associations with certain auditory and visual memory encounters from a film apparently seen many years ago. Cinema is used in his mind to deny the "enfleshing" of held hands especially by two men! Assuming he is heterosexual, he is self-creating a stereotype of gay men that serves to highlight and reinforce his particular form of white heterosexual perception. He is privileged to have that power over two film characters.

Alexis, however, has another take. He is Greek-Australian born in Australia—he sees beyond the constraints of his geography. "The Greek males do hold hands," he says. In Australia, however, he knows when, where, and how not to if necessary. This discussion on the call to flesh will be explored albeit lightly in the next part of the chapter. Meanwhile what happens to kissing?

Mairtin Mac an Ghaill, visiting Perth in 1998, spoke of the phenomenon of "rent boys." Their physical enactment of sexual expressions is a social code. In our discussion, Mac an Ghaill spoke of this groups' social code and how as an inquirer he wondered about their strata of behaving. As an ethnographer he was given the opportunity to be in their world, hearing and seeing the rituals of enactment he had heard of. The last thing that rent boys, "who were usually straight," would do was "kiss." If kissing ever did occur it was *never* in a public display of affection. Here we have another exemplar of the enfleshment that has been categorized in various ways. I would like to expand this a little more. First the rent boys use not kissing a "client" as a way of delineating a boundary of behaviors. The rent boys *know* that if they kiss in public they will be perceived as "poofs." Such a label through the act of the kiss is too much to entertain for these young men. Not kissing is an act of not connecting, thereby disempowering the "professional" relationship.

We *are each of us* immanent creatures and as such we only have this incarnation (*being made flesh*) to believe in and know ourselves. For those who can't or don't know the province of gay men, lesbian women, and intersex people the enfleshment of our being is *blatantly* denied. Susan, retelling her nine-year-old daughter and five-year-old son's discussion, provides a heteronormative viewpoint.

> My daughter and son were in the back of the car. He is 5 and she is 9. She was trying to tell him what being gay meant and all I wanted them to do was to pay attention to anything else—I even tried to divert their attention away from the subject by talking about sheep...I was really worried when they started to talk about what gays actually do. It was interesting that he could

cope with the boys doing things together but he seemed to be wondering how girls did it. He couldn't grasp that lesbians actually did physical things. My five-year-old son had already learned about the invisibility of the lesbian women. (Susan, a non-gay mother)

The apprehension Susan voices is softened as she hears her two children talking of physicalizing sensuality through lovemaking. Her comments indicate that the impressions that might be assembled from the loving of same gendered people for her daughter's viewpoint *are* well informed. The nine-year-old daughter says, "They can actually get married. But I don't think that they can really. They can love each other but they can and can't get married I think."

The capacity to be sexual is a principle of pleasure that children seem to already know about. The nine-year-old engages "marriage"—a supportive, regulatory, heterosexual ritual of great importance—to further illustrate the identity of the denounced gay men and lesbian women. Using marriage she brings to the discussion another enactment that illustrates her emerging awareness.

Often our own tales go like this: "*It is OK to be who you are but please don't tell me about it—'don't ask don't tell'—and please don't explain what you do in bed.*" The denial of flesh, the incarnate self is the reality for many intersexed people, for women who love women and *do* sexualize their passion; for men who choose to share their intimacies in the *art* of loving physically. The denial of the enfleshment is the meat of the debate that each of us lives with. As an open gay educator it is OK to "be" but not to "do" my sexual expression. As Peter McLaren confirms:

> ideological hegemony is not realised solely through the discursive mediations of the sociocultural order but through the enfleshment of unequal relationships of power. Hegemony is manifest intercorporeally through the actualization of the flesh and embedded in incarnate experience. (McLaren, 1991:169)

Thereby I am—and paradoxically I am not. Am I one who does hold hands or kiss my own gender? Or am I one who doesn't? Either way I get to live in a liminal space—a "forbidden zone." It depends not so much upon *my* need to touch another hand or lips with a kiss, but as on the viewers' configuration of what is right for me.

In discussing the theories of how the subjectivity of the body creates identity, McLaren emphasizes in his writings that there has been an overtextualization of reality. The word has come to mean the act. The word "gay" has come to mean the "act" of gay.[12]

Heterosexuals cannot appreciate gays unless they encounter gays. Textualized gays cannot be experienced or comprehended. They have become textualized even in this chapter and because of enforced invisibility their textual representation is skewed by the heteronormative world. Members of the heteronormative world have politicized their action toward gays by limiting the meaning that they give to the way of being for gays that is unique to the gay realm. If gays are seen as a fiction from the world of the heterosexuals, then their meaning configuration must be unbalanced. Heterosexuals know that real people exist but prefer to believe that "fairies live at the bottom of the garden."

Fairies at the Bottom of the Garden—a Heteronormative View

Fairies at the bottom of the garden have no power at all. Because they are fictionalized, they have no "form" or body. The fairy is created as required in the mind of heterosexuals who see the fairy in their own style through a limited field of vision. Heterosexuals have power to create the fairy in heterosexual terms because they don't look into the bottom of the garden to see the fairy. By implication the fairy becomes in the heterosexual field of knowing a reduced textual sign of what a fairy really is. How can the fairy at the bottom of the garden carry out political choices when the heterosexual people who "make" him actually create an "absent subject." Most of all, hetero-sexuals choose to see the fairy as having no body. The text creates no sense of what a fairy is for either the heterosexual or the fairy.

What is happening at the bottom of the garden that the heterosexual doesn't want to see? McLaren (1988) tells us that the body is a way to liberate us from the text and is a cultural artifact. So the fairy can and does use its time with other fairies to produce the fairies' own identities, through interactions of flesh which make meaning for *that* community. This interacting becomes a series of moments that are investments in the body and are what McLaren (1991:154–155) refers to as the "order of discourse." Because the fairies and the heterosexual people don't know each other, the "modes of desire," "modes of production," and "modes of subjectivity" all become somewhat obscured from each other. The modes of desire assumed by the fairies are declared by the heteronormative world to be wrong. The fairy body has become "a terrain of the flesh in which meaning is inscribed, constructed, and reconstituted" (McLaren, 1991:150). For both fairies and heterosexuals meaning has a bodily character.[13] Members of the heteronormative world are doomed to failure in attempting to constitute "flesh" through words and symbols as they cannot see real fairies.

Regrettably their impositioning language is not immaterial as can be seen from the narratives of the gay people in this chapter—heteronormative practice uses language to intensify and enlarge the politics of the fairies' bodies and flesh from only a *negative* viewpoint. The fairies they imagine and do not "know" become punching bags for heteronormative ignorance. In the main the emphasis on keeping the fairies invisible through "text only" has created an air of narrowmindedness, intolerance, and hatred toward gay men.

We are each of us alienated as we are being operated upon from a place of censorship in which as subjects we are forced to obey the norms governing what is speakable. Butler calls this "speakability." Wilde wrote of "the love that dares not speak its name." The consequence of "not speaking" for us is to be enforced to embody the norms of those that govern speakability and "to move outside the domain of speakability is to risk one's status" (Butler, 1997:133).

New research requires us to attempt to uncover ways to tell some of the hidden gender meanings beyond classical statements that still use the predominant psychological, social, and political philosophies to tell others who we are. These ways of understanding conspire to cause us to lose our voices (Harter, Waters and Whitesell, 1997). Current research into concepts of identity, and sexual identity especially, includes narrow samples, a prejudiced focus on sexuality, undeviating in nature as it is falsely formed by either the female or the male. The research seems to be characterized by a lack of attention to the larger sociohistorical and psychosocial contexts of those who deviate from the "either/or male-female" divide alluded to above. The consequence is the minoritizing of our, your and my, sexual identity. As "others," we are forced into the realm of silence or benign acceptance. The alienating of those who do not fit the linear samples of standard research is one premise underpinning this paper. The eventual cost we pay is denial of the self (Eliason, 1996). New inquiry must not only deconstruct the "textual" experience of the confining practices of our modern society. Research and inquiry must employ alternative ways to reinvent self-knowing—such as art making and other ritual practices briefly mentioned. Only through making visible the gendered others will we potentially reconstruct ourselves *in our own ways*. The act of becoming who we really are is the call of this chapter (Heilman and Goodman, 1996).

The hegemonic knowing only through the either/or (male-female) gaze and way of being is brought into question by the existence of other-gendered people. They are woven here in the chapters of this book. Western societies are raising the questions, but preferring not to hear answers, on what "they actually do." One way back to the quest to be considered as an authentic human being is the telling of our truth and our multiple meanings. To the heterocentric hegemony this brings insight and perhaps confusion. The enforced silencing of gendered others and the defleshing of their passion is painful, leading to psychosocial alienation and loss of self-hood (Gergen and Davis, 1997). The ideal is to be represented without repression. The denial of self is occasioned when the audiences of our lives deny that we are all human. We are not warped images of the psyches of the straight world, for each of us is in our own way real. The presentation of our realities is all that we are asking for—as do all individuals in their quest for completeness. Each of us has experienced the processes of repression leading to a dehydrated self; forced to not express our selfhood or worse become the object of voyeuristic spectators in lackluster drag shows in theaters of prostitution. This is no longer enough and times are changing and challenging.

References

Bornstein, K. (1994). *Gender outlaw: On men, women, and the rest of us.* New York: Routledge.
Butler, J. P. (1993). *Bodies that matter: On the discursive limits of "sex."* New York: Routledge.
Butler, J. P. (1997). *Excitable speech: A politics of the performative.* New York: Routledge.
Crosier, Louis, and Pat Bassett (eds.). (1994) Meeting the needs of gay and lesbian students: A plan of action for our schools. In *Looking ahead: Independent schools issues and answers.* Washington: Avocus Publications.
Eliason, M. J. (1996). Identity formation for lesbian, bisexual and gay persons: Beyond a "minoritizing" view. *Journal of Homosexuality,* 30:3.
Foucault, M. (1984). Politics and ethics: An interview. In *The Foucault reader,* ed. Paul Rabinow. New York: Pantheon.
Gergen, M. M., and S. N. Davis (eds.). (1997). *Toward a new psychology of gender.* New York: Routledge.
Goffman, E. (1974). *Frame analysis: An essay on the organization of experience.* New York: Harper & Row.
Goss, R. (1993). *Jesus acted up: A gay and lesbian manifesto.* San Francisco: Harper.
Harter, S., P. L. Waters, and N. R. Whitesell. (1997). Lack of voice as a manifestation of false self-behavior among adolescents: The school setting as a stage upon which the drama of authenticity is enacted. *Educational Psychologist,* 32:3.
Heilman, E., and J. Goodman. (1996). Teaching gender identity in high school. *High School Journal,* 79:3.
Lacan, J. (1977). *Ecrits: A selection.* Translated from the French by Alan Sheridan. London: Tavistock Publications.
Massachusetts Governor's commission on gay and lesbian youth. (1992). http://www.glsen.org/pages/sections/library/schooltools/033.article
McLaren, P. (1988). The liminal servant and the ritual roots of critical pedagogy.

Language Arts, 65:2.
McLaren, P. (1991). Schooling the postmodern body. In *Postmodernism, feminism, and cultural politics.* Ed. Henry A. Giroux. Albany, NY: State University of New York Press.
Rich, Adrienne. (1983). Compulsory heterosexuality and lesbian experience. In *Powers of desire: The politics of sexuality,* ed. Ann Snitow, Christen Stansell, and Sharon Thompson. New York: Monthly Review Press.
Slack, J. (1996). The theory and method of articulation in cultural studies. *Stuart Hall: Critical dialogues in cultural studies,* ed. D. Morley and K. H. Chen. New York: Routledge.
Snitow, Ann, Christen Stansell, and Sharon Thompson (eds.). (1983). *Powers of desire: The politics of sexuality.* New York: Monthly Review Press.

Notes

1. This chapter was originally part of a joint presentation with Delphine McFarlane and Chris Somers at the Third International Congress on Sex and Gender in Oxford, September 1988.
2. This is pretty much my understood experience of oppression and liberation. After hearing a radio talk on the anniversary of the International Women's Year I felt that women in their emancipation, which is still incomplete, had paved the way for others. The androgynous individual is so removed from the world (and invisible) that what the gay and lesbian political movements have done is pave the way for androgynous to be accepted.
3. Testimony of a 17-year-old lesbian student before the Massachusetts Governor's Commission on Gay and Lesbian Youth, November 1992 quoted in Crosier, Louis and Pat Bassett (eds.)(1994), Meeting the Needs of Gay and Lesbian Students: A Plan of Action for Our Schools. In *Looking Ahead: Independent Schools Issues and Answers.* Washington: Avocus Publications.
4. Poofters, fags, nancies, sissies, fairies, Homos, bent, Fruits, mince, big girls, queers, sodomites, Aunties, meat men, Rodents, She-males.
5. Butches, lezzos, dykes, Crack Snaker.
6. Gender benders, Post-ops, Pre-ops, Transvestites, Transsexuals, drag queens and kings, cross-dressers, dick-chicks.
7. I use this word because some heterosexual people would find the world gays live in insane—not of our own making. The sanity we need is based on respect, an ability to feel completely our selves. Any action that creates a sense of a person being "less than (<)" is just another "gram" of oppression that is insanity-making. Gay men are prone to the same feelings of worthiness as straight people, and also require similar occasions of recognition to construct and create their inner sense of sanity.
8. No doubt the reader is aware that gay bashing is still prevalent. The cause of this brutal activity is fear. The nature of the experience brings traumatic and unsafe feelings to the world of the victim. I intentionally write "victim." The assaulted persons are variously required to defend themselves from a premise of "guilt" in many agencies' eyes. For example, the comment that "he [the gay man] came

onto me" is frequently used as a rationalization for the attack. I wonder how many straight women kill or maim straight men in reply to the unwanted advances of the men. Gay bashing is real and hasn't gone away.

9. Is this simply the ability to be proud of who we really are? There seems to be a climate of change that is more positive toward gay people, though the Data Lounge, an American weekly gay/lesbian newsletter on http://www.data lounge.com/ indicates persistent unease with gay issues. The *Washington Post* (in association with the Henry J. Kaiser Family Foundation and Harvard University) concluded a national survey gauging American reactions to homosexuality. The results, say the newspaper, show that while attitudes toward gay men and women have changed dramatically, most people remain deeply conflicted and uncomfortable on the subject of gay acceptance....In the national poll, 87% said gay people should have equal rights in terms of job opportunities. Asked if sexual relations between consenting adults of any sex should be legalized, a much smaller majority, 55%, said yes, while 34% answered no.

10. In Western Australia, where these narratives largely originated, the age of consent between men is at variance with other states, and inconsistent with heterosexuals. Heterosexual males may have intercourse with heterosexual females at the age of 16. Gay men can have sex with gay men over 21 years of age. Additionally, the Western Australian government has not yet addressed anti-discrimination laws that can be applied to gay, lesbian, bisexual and transsexual (GLBT) people. In Western Australia a gay schoolteacher cannot openly speak about gay issues.

11. Being considered as an authentic human being is the telling of our truth and our multiple meanings.

12. The reader may be aware of the theological stance that you can be gay but should not "do" gay as a way to avoid wrathful sinfulness. The *Washington Post* article clearly indicates that the act of gay sex is what hampers the acceptance and tolerance of gays.

13. The Henry J. Kaiser Family Foundation and Harvard University research (see note 10) revealed that of US citizens polled, 50% still say homosexuality is unacceptable (29% say it is unacceptable, but should be tolerated by society; 28% say it is unacceptable and should not be tolerated). When the question asks about *gay sex*, opposition surges from 57% to 72%.

Gender Performativity and Normalizing Practices

Wayne Martino and Maria Pallotta-Chiarolli

> the production of social life is a *skilled* performance
> —Giddens in Cohen, *Structuration Theory*

Introduction
In this chapter we examine the normalizing regimes of practice that impact on the various ways in which young people at school define desirable forms of masculinity and femininity. Attention is given to the particular role that compulsory heterosexuality and gender duality play in prescribing appropriate behavior for boys and girls.

By drawing on interviews with adolescent boys, a 13-year-old self-defined "tomgirl," and an adult transgender woman, and examining written responses by girls, we highlight the kinds of issues that impact on their lives at school. We also consider the invisibility of transgender and intersexual perspectives in most educational debates on gender and sexuality.

Heteronormative and Gender-Normative Surveillance
Heteronormative currencies of masculinity and femininity frame the ways subjectivity is policed for adolescents through certain techniques and practices of self-regulation and surveillance in medical, educational, and everyday structures and practices. Foucault (1984a, 1984b) draws attention to the normalizing practices within which the limits of particular modes of subjectification are circumscribed. This has certain implications for understanding the norms governing the ways in which young people at school learn to fashion for themselves certain forms of masculinity and femininity. In this regard, Foucault highlights the pivotal role that sexuality has to play in the way that we learn to define ourselves and relate to ourselves and others:

> an effort [is made] to treat sexuality as the correlation of a domain of knowledge, *a type of normativity and a mode of relation to the self*; it means

trying to decipher how in Western societies, a complex experience is constituted from and around certain forms of behaviour: an experience which conjoins a field of study (*connaissance*) (with its own concepts, theories, diverse disciplines), a collection of rules (which differentiate the permissible from the forbidden, natural from monstrous, normal from pathological, what is decent from what is not, etc.), *a mode of relation between the individual and himself* (which enables him [sic] to recognise himself [sic] as a sexual subject amid others). (Foucault, 1984b: 333) (our emphasis)

It is in this sense that the specification of sexual behavior is tied to disciplinary regimens and apparatuses in which certain concepts, theories, and rules for governing conduct are formed according to an assemblage of historically contingent norms. This is important in any discussion of how young people in schools learn to fashion for themselves particular gendered identities with regard to the bodily enactment of such forms of subjectivity (see Butler, 1990; Nayak and Kehily, 1996; Epstein, 1997; Dixon, 1997; Connell, 1987, 1995). As Rajchman (1986:169) emphasizes:

It is a problem not simply in what we say about ourselves but in what we do to ourselves and our bodies. The constitution of the disciplined individual is the constitution of the disciplined body: of his [sic] soul or identity as the prison of his [sic] body. We can constitute ourselves by what we wear, where we live and what we eat. (our emphasis)

Our analysis of gender performativity in this chapter treats masculinities and femininities as an ensemble of self-fashioning practices that are linked to normalizing judgments and culturally specific techniques of the body and modes of thinking. We draw attention to how adolescents use the techniques and strategies available to them, within specific heteronormative regimes of practice, to constitute themselves as males and females of particular types. Moreover, in focusing on the ways in which they learn to relate to themselves, consideration is given to the practices of the self they adopt and apply in fashioning particular styles of masculinity and femininity.

This analysis is also in line with Mauss's (1973) position on techniques of the body and the production of desire. Mauss emphasizes that bodily and mental capacities are not naturally given, but are contingent upon social practices, cultural beliefs and the deployment of quite specific techniques that fall under the rubric of "miscellaneous social phenomena:" "I mean the ways in which from society to society men [sic] know how to use their bodies" (70). Mauss (1973:73), in fact, provides a descriptive analysis of the ways in which particular capacities and attributes such as styles of walking, running, swimming, sitting, and digging are bodily dispositions that are culturally acquired through "prestigious imitation" and specific techniques of social training.

> In all these elements of the art of using the human body, the facts of education were dominant. The notion of education could be superimposed on that of imitation. For there are particular children with very strong imitative faculties, others with very weak ones, but all of them go through the same education, such that we can understand the continuity of the concatenations. What takes place is prestigious imitation. The child, the adult, imitates actions which have succeeded and which he [sic] has seen successfully performed by people in whom he [sic] has confidence and who have authority over him [sic]. The action is imposed from without, from above, even if it is an exclusively biological action, involving his [sic] body. The individual borrows the series of movements which constitute it from the action executed in front of him [sic] or with him [sic] by others.

Such practices in "the art of using the human body" build through "education" and "imitation" a repertoire of skills and capacities which cannot be conceptualized as natural: "there is no 'natural way' for the adult—something we think of as normal, like giving birth lying on one's back, is no more normal than doing so in other positions, e.g. on all fours" (Mauss, 1973:74, 79).

Mauss's attention to the naturalization and normalization of the bodily enactment of subjectivity has implications for understanding the normalizing regimes of practice in which the performativity of gender and sexuality are imbricated. We are not always aware of the trainings to which we have been subjected which lead us to use our bodies in culturally specific and circumscribed ways. The formation of human attributes and capacities cannot be confined to social, biological, or psychological functions per se, but is the effect of an ensemble of practices in which all "three elements [are] indissolubly mixed...What emerges very clearly...is the fact that we are everywhere faced with physio-psychological-sociological assemblages of series of actions" (Mauss, 1973:74, 85).

Moreover, Mauss, in his classification of techniques of the body according to efficiency, highlights how such techniques and social forms of training are tied to a regime of normalizing practices. Norms are formed within specific social apparatuses and technologies through an ensemble of techniques and practices that make possible particular uses of the body and the formation of certain mental capacities (see also Hirst and Woolley, 1982; Hunter, 1991):

> *These techniques are thus human norms of human training. These procedures that we apply to animals, men [sic] voluntarily apply to*

themselves and to their children. The latter are probably the first beings to have been trained in this way, before all the animals, which first had to be tamed. As a result I could to a certain extent compare these techniques, them and their transmissions, to training systems, and rank them in the order of their effectiveness. (Mauss, 1973:78) (our emphasis).

It is important to note that Mauss conceptualizes human beings as *voluntarily* applying specific techniques to themselves and their children. He does not present trainings and techniques at the level of what Foucault (1980, 1991) would term *juridical-legal* or *sovereign*. In other words, techniques are not merely imposed upon individuals; they work on the self in such a way that the individual is located in a regime of self-regulation and monitoring, which leads to *voluntary* implication in specific procedures and techniques of the self (see also Foucault, 1985, 1986). Foucault (1977) uses the panopticon model as symbolic of the power of the wider social order in evoking self-regulatory practices. Sociocultural surveillance and self-regulation are interconnected:

> [The panopticon is] at once surveillance and observation, security and knowledge, individualization and totalization, isolation and transparency. (Rabinow, 1984:217)

Hierarchical, continuous, and functional surveillance is organized as a multiple, automatic, and anonymous power. This enables the disciplinary power to be everywhere and always alert, since it leaves no zone of shade. Disciplinary power is therefore exercised through its invisibility. At the same time, it imposes on those whom it controls a principle of compulsory visibility, which assures the hold of the power that is exercised over them. As Rabinow (1984:199) explains, it is the fact of being constantly seen, of being able always to be seen, that maintains the disciplined individual in his subjection.

It appears that each person is his/her own panopticon in the sense that he/she undertakes a particular policing and monitoring of the self, which is circumscribed by historically contingent norms (see also Kazmi, 1993).

Another way of theorizing the connections between sociocultural surveillance and self-regulation in relation to gender and sexuality in young people at school is via the metaphors of "performance" and "passing" (Butler, 1990; Anzaldua, 1990). These involve the presentation of a social identity or performance in a public "world" that may be out of harmony with a private identity or desire. As the research participants in this chapter illustrate, performance strategies include silencing, editing, imitating, masking, manufacturing, and parodying. For example, some students may construct and perform some form of alternative fictional gendered and sexual self in order to "pass" or they may selectively present/perform only certain aspects of their self to an audience of peers, teachers, and parents (Plummer, 1981). Other students may deliberately choose to transgress fixed hegemonic notions of gender and sexual duality and "play" with gender and sexual diversity, dealing with the resultant negative reactions. However, they can select their performances only from the options available within their sociocultural and socioeconomic contexts.

Butler's (1991) "performative theory" is enormously insightful as a framework for exploring the ongoing, interactive, imitative processes by means of which categories of the self, including gender and sexuality and their illusions of authenticity, are constructed. In her later text, Butler (1993:2) reformulates performativity not as the act by which a subject brings into being what s/he names, but, rather, as that "reiterative power of discourse to produce the phenomena that it regulates and constrains." Performativity acquires an act-like status, concealing or dissimulating the conventions of which it is a repetition.

At other points in her work, Butler's (1990) notion of gender parody does not assume that there is an oppositional original that such parodic identities imitate. As we will see in the gender performativity of Stephen later in this chapter, the parody transcends dualism as it is *of* the very notion of an original. As an imitation that effectively displaces the meaning of the original, it imitates the myth of originality itself. The various acts of gender create the idea of gender, and without those acts, there would be no gender at all. Paradoxically, the reconceptualization of identity as an *effect*, that is, as produced or generated, opens up possibilities of "agency" that are foreclosed by positions that see identity categories as foundational and fixed. Individuals such as Stephen and Jacqui Cussen can locate strategies of "subversive repetition" and "affirm the local possibilities of intervention through participating in precisely those practices of repetition that constitute identity and, therefore, present the immanent possibility of contesting them" (Butler, 1990:147). This type of "script manipulation" and passing means they not only monitor their own performance of the script, but also display a variety of styles in that monitoring. Script switching, script evasion, and juxtaposition are standard elements in many "dramatic models" (Cohen and Taylor, 1976:66). Anzaldua (1990:xv–xvi) also theorizes about "passing" and "performance" as forms of *haciendo caras* imposed upon one by the wider society but also providing skills for one to fashion one's own mask:

> *haciendo caras*, "making faces" means to put on a face...Some of us are forced to acquire the ability, like a chameleon, to change color when the dangers are many and the options few... Between the masks we've internalized, one on top of another, are our interfaces...between the masks that provide the space from which we can thrust out and crack the masks...We begin to acquire the agency of making our own *caras*.

The above theoretical discussion and resultant questions regarding young people at school provide a necessary deconstructive and problematizing framework for the ongoing debates surrounding what psychiatry and psychotherapy have constructed and labeled as Gender Identity Dysphoria (GID) in children and adolescents. Practitioners and clinicians such as Coates (1987) and Zucker and Bradley (1995) utilize gender-normative and heteronormative constructs of identity to rationalize and justify psychiatric and other clinical interventions with children and adolescents who exhibit what is binarily constructed as "cross-gender identification." The focus of these therapeutic interventions is the reinstatement of a "natural" gender duality by pathologizing the child and his/her lack of conformity to a panopticonic society's hegemonic norms for fashioning a feminine or masculine heterosexual self.

The school is a major site of this naturalizing of gender duality via regulation and surveillance. Any difficulties and harassment experienced at school by children and adolescents who do not conform to fixed normative notions of masculinity and femininity are often used to justify therapeutic intervention to modify the child rather than educational interventions to modify the school. The following rationales for why GID has been diagnosed and defined as a condition and why it needs to be treated exemplify the use of "natural" and "normal" sociocultural constructions of fixed gender duality and heterosexuality, and the *acceptance of* rather than *resistance to* the panopticonic social systems of regulation and surveillance. Indeed, the construction of GID and its consequent medical infrastructure for treatment is another form of panopticonic surveillance and regulation. It instructs parents and teachers on how to detect and diagnose GID, and sets up the expectation that as part of their role as efficient and caring adults, they will seek out and support the "curing" of the child for the good of the child. Otherwise, they can be labelled and punished as negligent and abnormal.

> At least three goals—elimination of peer ostracism in childhood, treatment of other psychopathology, and prevention of transsexualism in adulthood—are so obviously clinically valid and consistent with the ethics

of our time that they constitute sufficient justification for therapeutic intervention (Zucker and Bradley, 1995:269)

> These kids are getting a lot of flak from their peers, and that causes a lot of stress...We don't know how to change society, but we can change their gender identity problems so they can live in their peer group with less distress. (Bahlburg, quoted in Minter, 1999:18)

> The gender-disturbed child moving into adolescence is a sensitive individual with reduced anxiety to tolerance, a rather weak sense of himself or herself, and a degree of gender insecurity. These factors may make it very difficult for this individual to contemplate heterosexual involvement...Once begun,...sexual activity in itself is reinforcing because of the pleasure involved and the self-validation, both from other partners and from a group. This produces...a homosexual identity. (Bradley, quoted in Minter, 1999: 21)

Some educators and medical practitioners criticize the construction and deployment of GID:

> Is psychiatry, with the diagnosis of GID and diagnostic criterion of gender dysphoria, simply recreating a stigmatization that we experienced with homosexuality in the fifties and sixties?...If our conception of gender were more fluid, would not the very notion of gender "nonconformity" be nonsensical?...Differently gendered lives—their individual variation, their difference from the majority—constitute a *normal diversity* of gendered experience. (Rottnek, 1999:5–6)

Minter (1999:27) also points to the heterosexism framing the identification and treatment of GID:

> if GID in children was not strongly associated with homosexuality in adulthood, it is unlikely that "feminine' behavior in boys and "masculine" behavior in girls would have been designated psychiatric disorders or become the focus of an entire clinical field devoted to analyzing and "correcting" cross-gender behaviors and identifications... attempts to alter or manipulate a child's future sexual orientation and/or gender identity raise serious ethical and legal concerns.

As Neisen (1992:65–66) argues:

> I was appalled...[that] research is still being conducted to differentiate between so called "normal" and "gender-disturbed" boys...the victim gets blamed while the perpetrator goes unchallenged. Implicit is that feminine behavior in itself is devalued...so too are there heterosexist overtones.

In the rest of this chapter, we will consider the experiences of young people in our research in relation to the above theories and debates. The following questions will frame our analysis:

- How and why do adolescents fashion for themselves a particular form of

subjectivity?
- What practices do they engage in to decipher "who they are"?
- How are certain desires formed within a regime of knowledge-power relations that serve as an index of their masculinity or femininity? In other words, what enables young people at school to recognize themselves as proper incumbents of certain gender categories?
- What happens when students transgress or cross the gender norms prescribed for enacting valorized heteronormative gender categories?

Fashioning and Performing the Gendered Self

> We're not born knowing what society expects from a male or female to be like. It's something we learn from day one.
> —Heather Smith, 16 years old, "Following the Gender Rules"

The call to girls for submissions for *Girls Talk: Young Women Speak Their Hearts and Minds* (Pallotta-Chiarolli, 1998) was based on the premise that girls have much to teach about the policing and surveillance of gender diversity and sexual diversity in their lives. Through their writings on issues such as body image, schooling, friendships, sexuality, and sexual relationships, many girls discussed the impact of heteronormativity and gender performativity on their self-regulatory and self-fashioning practices. For example, Heather Smith describes her growing realization that imitation of her father's behavior and bodily exposure began to conflict with the need to fit into "normal" feminine bodily exposure and behavior:

> on hot days when I was about three or four and I was with Dad, I just did as he did; if he took his shirt off, I did the same. But as I grew older, society and the media told me what I did was not what "girls do," so I stopped and considered everything I did. Is taking your shirt off ladylike? (Smith, 1998:1–2)

In relation to the theorization of Foucault and Mauss discussed earlier, here is Heather's analysis of panopticonic surveillance and self-regulation based on gender normativity and heteronormativity:

> Anybody who disobeys these rules is seen as a "genderbender," "poof," "lemon," all meant as put-downs. So as adolescents we strive very hard to play these male and female roles that are defined for us. Being accepted is a major part of teenage life, and we will compromise our feelings just to fit in. (Smith, 1998:1–2)

Heather concludes, "We need to break the 'female moulds' and 'male moulds' that society has defined and make our own individual mould" (1998:4).

Twelve-year-old Bethwyn Miller's description of her favorite activities reveals the conflict between normalization practices and resistant agency. Although she describes the activities she enjoys, she follows this with a self-regulatory denial of masculinity or sexual abnormality:

> On the weekends, I like mucking around with my tools. I get pieces of wood and saw the ends off, hammer nails into it, sand it down etc. My brother Rohan and I sometimes get my football and kick it to each other along the street. It's a lot of fun. After school on Tuesday, I go to karate. But all this doesn't mean I am an all-masculine tomboy. I am a girl and proud of it. (Miller, 1998:12)

Here are three more examples of girls' awareness of the gender-normative and heteronormative regulations defining their body shapes, movements, and adornment:

> It bothers me that it is rude for girls to sit with their legs open in a more relaxed way but fine for boys. (Catherine Pendrey, 10 years old, 1998:12)
>
> If a girl's not developed or something, you know, they get a lot of teasing, yeah, from the guys. (Kate, 16 years old, in Martino, 1998a:146)
>
> I want to shave my head and pierce my nose. I want to buy a big jacket and some work boots. I'm really sick of the drudgery and conformity. I suppose shaving my head would sort of reinforce my sexuality, and I'd get heaps of flack for it. I feel like going all or nothing. I'm sick of this half-halfness of my current life. (Jess Langley, 1998:220)

Thus, the work of Foucault and Mauss clearly has implications for theorizing the ways in which young people use their bodies as instruments within specific regimes of practice. "What specific forms of social training confer certain bodily capacities and skills within particular gendered regimes of practice?" is a question that is informed by Mauss's theorization of the link between social techniques and the formation of bodily capacities and skills.

This raises questions about the ways in which children and adolescents learn to walk, talk, and use their bodies in very specific ways. An interview with Stephen, aged 13, who identifies as a "tomgirl," exemplifies how posture and particular styles of using one's body and performing gender on specific occasions must be conceptualized as socially acquired bodily attributes and "prestigious imitation" that have been developed through the child's involvement in a range and ensemble of social practices (Mauss, 1973; Butler, 1990). According to clinicians such as Coates (1987), Zucker and Bradley (1995) and others, Stephen would be classified as exhibiting symptoms of GID.

Stephen: Because I always dress up like a girl, I'm always playing with mum's make up, I feel different from all the other boys, like I'm always dressing up in girls' clothes, do things that girls do. I used to be into Barbie dolls too...

Maria: And when did you first come up with the word "tomgirl" to describe yourself?

Stephen: Last year, probably.

Maria: Where did you get it from?

Stephen: I just got it from kids because girls who dress up like and act like boys and stuff, they call them tomboys, and I've heard teachers saying it, so guys who dress like girls are tomgirls.

Rather than labelling himself as boy or girl, Stephen self-identifies as a "tomgirl" after experiencing four interactive processes: external classification by peers and teachers; external surveillance by peers and teachers; panopticonic self-fashioning; and transgressive agency in refusing to perform or pass as either girl or boy, but devising a new "tomgirl" self based on Stephen's imitation and manipulation of the available sociocultural and populist options of masculinity and femininity.

During the interview, Stephen stood up several times to demonstrate to the interviewer how the walk and other bodily styles that were components of being a "tomgirl" had been fashioned. Stephen explained these were based on the observation and imitation of girls at school and women in society, and on men s/he had seen in magazines and on television performing femininity as "drag queens":

Stephen: When I'm wearing girls' clothes I always walk around [stands up gracefully], strut [begins to walk slowly with hips forward], like wiggling my butt [does so] and walk like that [swinging arms and head slowly and sensually], sort of curvy [accentuating hip-swaying].

Maria: So, when you walk as a boy?

Stephen: I just walk normally [ceases all swaying and accentuated head, arm and hip-movement], like that [walking straight ahead quickly], walking up and down [keeps body rigid, taking larger steps, louder stomping on the floor], straight [head forward slightly, shoulders hunched, hands as fists].

Maria: Which walk do you prefer?

Stephen: Bobbing up and down like that.

This last walk was a mixture of both the gender-normative masculine and feminine styles Stephen had just performed, as well as introducing a new up-and-down easygoing rhythm. Thus, we can see how Stephen has imitated and fashioned a self that can perform, indeed parody and imitate, a gender according to social context and

Gender Performativity and Normalizing Practices 97

individual desire, as well as devising a walk that transgresses both. Nevertheless, it is also interesting that Stephen labels the masculine walk as "normal."

Stephen also talked about a close friend at school, a "tomboy," who had fashioned a self that imitated and performed a hegemonic construction of aggressive masculinity:

Maria: Why do you call her a tomboy?
Stephen: Because she walks like a boy, she dresses like a boy, she looks like a boy sometimes.
Maria: Does she get picked on by other people?
Stephen: No, not really, she's got a bad temper. Like, if you say something to her she'll be at you, she'll hit you or something...you can't say anything to her because she'll get really angry, she'll push tables, chairs, she'll get you.

As well as his bodily movements, Stephen also utilized clothing to distinguish between and perform various forms of masculinities and femininities, based largely on styles popularized in teen popular culture. For instance, despite ongoing verbal harassment from some male peers, Stephen had fashioned him/herself as Mel B from the Spice Girls for the week during his/her thirteenth birthday, including makeup worn to his/her "Spice Girls Birthday Party." More recently, Stephen was fashioning various forms of masculinity by adopting various masculine dress codes. During the interview, Stephen was wearing what s/he described as "surfie shirt, baggy shorts, and sneakers":

> Kids always pick on me when I wear something different, because I think it was last week I wanted to be a punk,...I had the army pants, the jacket, the boots, the spiky hair...they were teasing me about it. And then I wanted to be a dance party guy ...You know, I've got the dance pants and the shirts, they've got little lights, different coloured lights, and the pants glow and...I had them on last week...I wanted to be one of them, kind of thing, but now I don't. Now I want to be a surfer and then I don't know, I could change next week or something.

Maria: So, what makes it keep changing?
Stephen: I don't know, I might see someone, I just like different varieties of clothes.

During the interview, Stephen spoke positively of the support s/he received from his teachers and some of his peers. Nevertheless, there were school structures that restricted his/her embodiment of gender diversity with their insistence on gender duality:

Stephen: We had to line up in our classes, girls in one row and boys in the other, it's so sexist...

Maria: How did you feel when you were in two lines, where would you have wanted to be?
Stephen: I wanted to be in the girls' line.

As this section has illustrated, a focus on bodily techniques and capacities is useful in elaborating a theorization of masculinities and femininities that attempts to account for the various ways in which adolescents establish a "stylized" regulated masculinity and femininity or fashion a transgressive performance of gender/s within a regime of normalizing practices.

Fashioning and Performing the Heteronormative Self

In this section we draw attention to the pivotal role that heteronormativity plays in the ways that adolescents learn to fashion for themselves particular forms of masculinity and femininity. We focus on gender performativity by referring to self-fashioning practices that constitute straight acting behaviors and, hence, draw attention to the pivotal role that sexuality plays in policing particular forms of masculinity (Epstein, 1997; Redman, 1996; Frank, 1993; Martino, 1998b; Ward, 1995; Butler, 1996; Nayak and Kehily, 1996; Kehily and Nayak, 1997; Mac an Ghaill, 1994; Connell, 1992). We argue that the anxieties some students experience with regard to establishing and maintaining a desirable form of masculinity and femininity, and which are then vulnerable to diagnosis and pathologizing as GID, are linked to techniques of self-problematization within a particular heterosexual economy of normative rules and codes of behavior (Rottnek, 1999). It is this fear and rejection of the *feminized other or masculinized other* that often governs the homophobic and heterosexist policing of desirable masculinity and femininity for many adolescents. What we want to focus on here, however, is the bodily enactment of such versions of masculinity and femininity. Through an ensemble of normalizing social practices and rules governing gendered conduct, students appear to have acquired a specific repertoire of skills and bodily capacities for establishing desirable forms of masculinity and femininity.

Scott, aged 16, talks about the footballer/surfie kind of guys who wielded a lot of power at the Catholic coeducational school he attended. They were the "cool" boys, the straight boys who

> just sort of get by hassling everybody and just having a few people they turn to and get them to laugh at the other people and it's all a bit of a mess really...They walk past and hassle all these other groups and make comments about them..."Oh I hate him, he's really gay" "Gee, those guys are real whatever." I mean it always gets back to gay with all these guys. It's

sort of like the big insult.
Wayne: They call other people "gay" do they?
Scott: Yeah, and we've had one guy right through from Year 8 and he got stuck with it and everyone decided, oh, you know, oh *he's definitely, he's a poofter, we hate him. You can tell the way he talks, he's friends with a lot of girls more than guys.* I mean, it's his choice really, but everyone's sort of stuck him with that tag and what surprises me when you talk to him, he's actually really homophobic himself.

Here Scott highlights how sexuality is used to define appropriate masculinity. He refers to one boy who has been identified as a "poofter." The criteria for labeling him in this way relate to his manner of speaking and that he tends to associate with girls as friends.

This boy, therefore, is targeted on two counts, both of which have associations with the "feminine" and a rejection of a particular kind of masculinity (see Jordan, 1995; Frank, 1993). The way this boy talks, presumably, is considered to be outside the range of acceptable modes of speaking for males, and his preference for having girls as friends also confirms that he is defying the norms. For example, boys are expected to talk in a particular way and to "do masculinity" with other boys. If a boy at school chooses to spend time with girls as friends, he risks having his masculinity brought into question; he may risk becoming a target of homophobic harassment (Martino, 1998c).

Researchers such as Hite (1981) have highlighted the role of fathers and other males in instructing boys not to be "sissies" and not to associate with girls as they are growing up. Such practices or ways of relating immediately become more visible in schools where any difference in behavior is readily recognizable in the open space of the playground, sportsfield or classroom. Within such spaces students are not only monitored by school personnel, but learn to monitor themselves according to particular norms and expectations that involve issues of sexuality. This is reflected in the comment that Scott makes about the boy who is targeted as gay actually being homophobic himself, pointing to the extent to which this boy has learned to monitor and regulate himself according to the dictates of compulsory heterosexuality.

Interviews with lesbian students conducted by Michelle Rogers (1998) also reveal their awareness and negotiation of surveillance from teachers and other students. This may involve the interrogation of fixed labels of the gay/straight binary divide that some school policies may be using to structure all anti-homophobic work and yet that also do not encompass the broader dimensions of sexual diversity. For example, Cloe (in Rogers, 1998:71) says,

> It doesn't really bug me who I am in love with as long as I am happy. I don't think any boxes can be put onto it [sexuality]. I don't really go, "OK, I am sleeping with a girl so I am gay," you know, I go, "I am totally in love with this person and that's what it means and that's what matters."

Young lesbian students critically question institutional rhetoric as they are aware of the contradiction between its purporting to support diversity while policing this diversity through silence, denial, and verbal, physical and emotional harassment, as well as endorsing self-regulatory practices. As Lisa says: "I know one of our mottoes of our school is like 'enabling you to find out who you are' but I'm not sure how the school would react if someone was very openly 'out'" (in Rogers, 1998:139). Similarly, Helena (in Rogers, 1998:97, 140) writes:

> They all go, "This is a heterosexual school and you shouldn't be here."
> I don't see a name on the wall saying this is a heterosexual school.

The following writing illustrates the heteronormative surveillance that not only enforces heterosexuality but also the oppositional fixed category of homosexuality. In this way, transgressors can be neatly categorized and easily contained and managed.

> she was a girl and I was supposed to like boys. At the time I rationalised that what I felt for her was pure admiration. Like so many others, I gracefully accepted what I thought to be my place in the world....Throughout my later teens I found myself in various relationships with guys. (Belinda Pursey, 1998:67–68)

Heteronormative surveillance and regulation means one can be heterosexual and still not be seen to conform to the specifications of a heteronormative society. Sally, aged 13 (Sally, Lacey, and Grandma, 1998), clearly understands that "the problem is not having a lesbian mum but having a lesbian mum in this homophobic world" (106). She has developed what she calls "categories, the people who know and the people who don't know," slotting people into these categories in order to assist her in negotiating her location in the sites of school, home, and small rural community in New South Wales (107). "Friends that meet you [Mum] say 'You've got a nice mum.' But if I told them you're a lesbian, they'd just change everything they know or think about you" (106).

Even school structures that have been established to supposedly deal with student ostracism and isolation due to homophobia are seen as complicit in the harassment as they are themselves positioned and policed by a heteronormative framework: "you're supposed to go to them [school counselors] with problems. But is it really, like, a problem? Imagine saying, "Hello, my mum's a lesbian." What's the problem?" (107). In the above statement, Sally directly challenges and resists the problematization of the child in therapeutic interventions rather than the problematization of the school's social order (Neisen, 1992).

Dave, an adolescent boy, talked at length to us about the homophobic harassment he received at his previous school, a Catholic single sex boys' school, at the hands of a particular group of boys:

> Dave: Yeah, they targeted me, another guy, there might have been a couple of others, but it was largely associated with the fact that a lot of the things we did were perhaps stereotypical of what they saw as being related to homosexuality.
> Wayne: Like?
> Dave: Like at the time I did ballet, and the guy I knew, his posture and manner and everything he did was supposedly leading towards something that they suspect that he would turn out homosexual. So, they abused him for that, and me for that. I remember a couple of others, I can't remember what they did, but I remember distinctly that they were harassed for being homosexual. Well, they weren't, but they were seen as homosexual, whether they had a girlfriend or not, it was just that they paid no attention to sort of the facts.

Here attention is drawn to a regime of normalizing practices in which homophobic strategies are used against boys who are identified as homosexual on the basis of their posturing, manner, or involvement in ballet. In fact these are the very diagnostic strategies that are deployed by practitioners identifying GID (Coates, 1987; see also Minter, 1999). Thus, there are certain mannerisms and performative practices involving body posturing, as well as the kind of voice a boy has, which lead particular boys to be categorized as gay or as experiencing gender dysphoria (see Walker, 1988; Nayak and Kehily, 1996). Furthermore, such practices are imbricated in the way that many boys learn to establish their masculinities, at the level of performativity, through processes of differentiation in which indicators of homosexuality are readily recognizable as markers of deviance from a heterosexual norm, and hence from an appropriate masculinity (see Epstein, 1994; Steinberg et al., 1997; Redman, 1996; Willis, 1977; Connell, 1987, 1995).

Dave highlights in his interview an apparent "culture" of violence that was officially endorsed and socially sanctioned in the single sex Catholic boys' school he attended:

> Dave: I remember from, say, Year 5 that there were these every strong, very

> anti-homosexual feelings, everyone was very much against it. Maybe because, knowing that in the school, at this private Catholic school, the all-boys' school, it was felt that obviously it was not socially acceptable and that it was very bad to feel like that. If you were a homosexual, it was very bad, it was against what society is meant to be. The values that they experienced, it said that homosexuals were just a lower form.
>
> Wayne: So how did they know if someone was homosexual?
>
> Dave: No, they don't. They assume, if they like the person, they won't press it, if they don't like the person, which is usually the case, that is if they show any sign of weakness or compassion then other people jump to conclusions and bring them down. So really it's a survival of the fittest. It's not very good to be sensitive. If you have feeling and compassion or anything like that, you wouldn't survive in a place like that.

Attention is drawn here to a regime of bullying practices in which those boys who are sensitive or who display "any sign of weakness" risk becoming targets of homophobic violence at the hands of a "pack" of other boys who acquire and maintain a status at the top of a pecking order of masculinities (see Connell, 1987, 1995). This is highlighted in Dave's reference to the "survival of the fittest," which signals that he is drawing on a particular body of gendered knowledge grounded in a form of biological determinism to account for the behavior of his peers. Such behaviors, which involve enacting a form of power, are interpreted, in this particular case, as instances of boys publicly enacting and performing a stylized form of heterosexual masculinity (see Nayak and Kehily, 1996; Haywood, 1993; Parker, 1996; Epstein, 1997; Dixon, 1997; Redman, 1996; Skelton, 1997). Within such a regime, the requirement to deny feeling and sensitivity—to act tough—is enforced. Those boys who visibly exhibit behaviors imputed to homosexuals, or engage in practices deemed to be sex-inappropriate for boys, risk being targeted and harassed at school.

Eventually Dave's mother withdrew him from the boys' school and enrolled him at a Catholic coeducational school where he still continued to experience homophobic harassment at the hands of a group of dominant "footballer, surfie kind of guys." What is important to note is that boys from this group were friends with boys from Dave's previous school and word had spread about his reputation before he had even arrived at his new school.

> Dave: It was again for my dancing. A lot of stuff like "ballet boy." I don't know, they were very unimaginative. Just a lot to do with being a woman, being homosexual was a big thing again because of dancing.

Gender Performativity and Normalizing Practices 103

Wayne: Who was doing that here?
Dave: It was a large group at that time in Year 9...I was condemned by that big group of footballers on the oval.

Here Dave highlights the role of normalizing practices that involve designating ballet or dance as a feminized activity. Once again, the policing of heterosexual masculinities is framed in terms of identifying sex-inappropriate practices, which then form the basis for imputing homosexuality. It is not so much dance "itself" that poses a problem, but its *association* with the "feminine." This link is produced through a regime of practices that are imbricated in regulatory technologies of the gendered self (see Laskey and Beavis, 1996; Mason and Tomsen, 1997; Steinberg et al., 1997; Parker, 1996; Haywood, 1993).

Wayne: So how do you see that group on the oval?
Dave: I'd see them as a group that thinks themselves the most popular.
Wayne: Popular? How are they popular?
Dave: Socially acceptable, I think, compared to the other groups whom they see as maybe inferior to them in their social acceptability.
Wayne: What makes them [the footballers] popular, do you think?...
Dave: I think it was for the boys their *masculinity*. They had a lot of football players, fights, threats, male attitudes were very much bolstered by each other. They kept each other going. So you had practically everyone in that group doing football, drinking beer, smoking, anything rebellious yet within the lines ruled by society as acceptable. They also talked about the women they had...they believed they had good looks...they ignored other people. I think this accounts for the whole group; they ignored others who were not in their rank and they always kept to themselves. They considered themselves good looking and had a lot of girlfriend/boyfriend relationships and having sex was big talk. The younger you were when you had it the first time, the better. It seems pretty primitive, but that was very big for them.

What are interesting are the normative ties that Dave establishes between the category of "masculinity" and a range of social practices. For Dave, displaying masculinity for these boys is linked to asserting their heterosexuality publicly by boasting about their sexual exploits with girls (see Holland et al., 1993; Willis, 1977; Hite, 1981; Wood, 1984; Walker, 1988; Kehily and Nayak, 1997; Haywood, 1993; Frank, 1993). Dave also highlights the extent to which these boys bolstered such a stylized form of heterosexual masculinity through social practices that involved playing football, smoking, and drinking. On the basis of engaging in such practices and "having good looks" these boys were perceived to acquire a high status masculinity.

The desirability of enacting such a form of heterosexual masculinity is further emphasized when Dave is asked to explain what he means when he describes these boys as *displaying their masculinity*:

> I'm not sure, but what I see as a *masculine* attitude is that they value certain things like bodily physical strength and attractiveness, like they have to be physically attractive to the opposite sex as in they have to be very strong, handsome, charming...able to get [the] attention of the opposite sex readily and easily whenever they wanted [*sic*]. They also have to be sport orientated, very sporty, very fit, able to do any sport and do it well. Intelligent, well it's not always good to not have a brain so they definitely want a bit of intelligence but they don't put too much emphasis on being brainy; it's more like not being too thick or stupid.

Dave's comments can be deployed to draw attention to the stylised demeanor of the "footballers." The requirements for displaying a particular heterosexual masculinity are spelled out in terms of demonstrating physical strength, being able to attract the opposite sex readily and engaging actively in a range of sports. Moreover, Dave claims that boys need not to appear too stupid, while at the same time avoiding the risk of presenting themselves as too intelligent, because both positions contravene the normalizing boundaries within which a high status masculinity is enacted.

What needs to be highlighted is the role that heteronormativity plays in the lives of these boys at school. In fact, the performative dimension of such self-fashioning practices for bodily enacting straight masculinities is highlighted in the urban gay social context where an equally oppressive regime of masculinity operates (see Signorile, 1997). Acting like a straight guy is perceived to be "normal" and desirable and is set against the stereotypes of the raging "queen" who dresses in drag or the feminized camp "poof" who waves his wrist around! "Acting straight" is perceived to be a desirable practice for many gay men and appears to be organized around norms that designate gender behavior in stereotypically oppositional terms with masculinity at one end of the continuum being defined in relation to femininity at the other. In other words, it is about narrowly defining masculinity in terms that are borrowed from within a regime of heteronormative practices for fashioning gendered subjectivity. This is highlighted by Jason, a young gay man, aged 20, who reflects on his life at school and currently at university. He starts to talk about how he has learned to fashion himself in ways that do not lead others to mark him as visibly gay-acting:

> I'm out to people who ask but I don't sort of wear stereotyped clothes

and have a limp wrist and all that. So I think a lot of homophobia is directed towards people who look gay, I mean they may not even be gay. Whereas I'm not like that...So I guess I'm not in a position where I might experience that as much. Because if I go into a shop that's full of strangers they don't think, "Oh he's gay."...I think it's just about being visible in some way as being gay. The silly thing is I guess that a lot of people, you see this at Uni [the University] in particular, people have images. For example there's the art student image and that tends to merge with a kind of gay image and you think, they look gay. I've got the stereotype of what gay is as well, as much as any straight person has. And you see them and you think, oh, they look gay. But they're not. You've either heard of them or you know of them, see them with their girlfriend. So, whereas a lot of my friends who are gay are what you'd call straight-acting, you just would not know they're gay, they're, I don't know, masculine I guess is what they are...But I think you don't suffer as much if you're straight acting, because you're just not visible. So that's what it's about, it's like if a black person could look white, that sort of sounds stupid, but they wouldn't get hassled for being black. It's a silly analogy but you know that's how it is.

Jason draws attention to how he performs gender and how it is possible for those who take on the demeanor or *haciendo caras* of straight-acting masculinities to "pass," to be not easily identifiable as gay and, therefore, not as visible and subject to homophobia. When we ask Jason to elaborate on his definition of straight-acting he says:

Jason: Well it's how you dress. If you dress, yes, in a way which is, it's hard to define isn't it? If your dress is more sort of simple, straightforward...And it's just your actions, you don't put it around that you're gay or anything. I mean that sounds silly, it sounds like non-straight-acting people do try and be flamboyant or something; they probably don't, it's just how they are. I think it's just being masculine.

Wayne: So how does someone be masculine? I'm interested in that.

Jason: Well that's a good point, I think it's how you talk, it's bodily movements, demeanour I guess generally. But it's an interesting question. I think it's just mainly body language and how you talk. I don't think it's anything more than that. That's how I've always sort of thought of me as being straight acting. I don't like the term, it sort of suggests you're acting, you're not but it's like you are.

Wayne: So it's like it's a kind of façade or something, is that why you don't like it? Or it's something dishonest about it that you're inventing it. And that kind of bothers you in a way?

Jason: Yes, not in a huge way. I mean every now and then you'll see someone and you make a point about it or something, often a very non-straight-acting person...they're not necessarily explicitly saying you're not being who you are, and it's ridiculous. As if they are being who they are, and I'm not. I should be like that if I were gay. I mean, I think gay people have just got all these stereotypes in their head just like straight people, about what it is to be gay, or straight. And we use them probably just as much.

Here Jason highlights the kind of norms prescribing specific enactment of gay/straight masculinities and draws attention to a form of policing and self-regulation that occurs in the homosocial gay context. He indicates that he feels pressured to conform to certain stereotypes about how gay men should conduct and bodily enact their masculinity and highlights the point that such practices are about the bodily movements, body language, demeanor and the way you talk!

Likewise, Tamsin Dancer (1998;157) discusses her transgression of homonormative expectations of women by identifying as bisexual rather than lesbian: "it means I don't make the grade in the eyes of many young dykes. When I came out, I knew if I told a lesbian I was bisexual they'd probably reject me, so I decided to come out as a lesbian."

One relationship with a man resulted in ostracism: "I felt as though I was playing a board game and had turned up a card that read 'You realise you are bisexual. Take three steps back (into the closet)," (Dancer, 1998:108).

Dancer is also critical of the policing of body and behavior within a homosocial context:

> I continue to be totally intimidated by "scene queens." I aspire to their looks—they are all slick and slim, and so very cool. But why should I use them as my measuring stick of normalcy? Why should they hold the status? We need to question why the chic-looking, white and well-dressed lesbian image is attractive. There are a lot of young dykes, bi and queer women, like me, that will never fit that limited category—we are fat, or poor, or people of colour, or parents or disabled. (158)

Scott, aged 15, also alludes to the specifiable limits of a particular regime of normalizing practices and compulsory heterosexuality. He draws attention, however, to the way that such limits are marked out by parents in terms of designating the kind of clothes that "you're allowed to wear." In fact, he claims that boys tend to be more restricted than girls in terms of what they can wear:

> Scott: I suppose it depends on your actual upbringing, your parents and stuff, but what starts off is like the clothes you wear and you're like allowed to wear. Because I know my parents sort of...I don't know...parents sort of want you to....they don't want their kids to turn out funny or anything. So they don't want them sort of wearing girls' clothes or wearing boys'—I suppose girls wear boys' sort of clothes and it's more acceptable for a girl to have short hair than a

guy to have long hair and certainly you don't see guys walking around in dresses and stuff without people sort of looking at them really strangely. Just that whole thing, whereas a girl can wear jeans and parents have no problem. So I suppose girls on one level have it easier than the guys in that idea, but then there's like the whole pregnancy thing where a girl's got it worse than the guy because the guy's parents don't have to find out I suppose, as necessarily. Not that they don't necessarily have to find out, whereas the girl's family's got to know about it, and that sort of thing. But I don't really have any experience of that sort of situation...only like clothes or whatever.

Wayne: So you're saying in terms of clothes guys are more restricted?

Scott: Yeah, like, if a guy dyes his hair or something, that's happened with me, everyone's going, "Oh no! He must be a queer person." It's died down now, but people still bring it up to me and stuff. It doesn't bother me.

Wayne: Is this at home as well, or is it more so outside?

Scott: More so outside, but I know my mum sort of gets a bit annoyed.

Wayne: Annoyed?

Scott: Annoyed with me sort of not caring... like you know, if I do girlish sort of things ... um ... I don't know, that whole idea they don't want to have kids being sort of funny, don't want a funny, queer sort of kid or something. *(He holds up his finger with red nail polish painted on it.)* I know like um that's really annoyed my mum 'cause I did that.

Wayne: The nail polish on your finger?

Scott: Yeah, mum got annoyed about that. Sort of strange really, 'cause I can't see it means anything.

Wayne: Why do you think she got annoyed?

Scott: Um, I dunno.

Wayne: You got nail polish on one finger, is that all? Just one finger?

Scott: Um, on my foot as well, but I mean Mum just got really annoyed, so I took it off my foot, but I left that *(holds up his finger)*. I don't know ...whatever...it's just the way my parents are.

Scott indicates the regimes of normalizing practices at home in terms of the role his parents play in policing a particular form of masculinity by attempting to regulate what he wears and how he chooses to present himself. The reason for this, he adds, is that parents don't want their children to "turn out funny." In short, he is required to display himself as appropriately masculine in terms of the way he dresses, but he decides to transgress these norms. The regulatory behavior of his parents, which is implicated in regimes of compulsory heterosexuality, however, may also be related to concerns about their son "acting differently" for fear that he will become a target of bullying practices. As stated earlier, it is these concerns that encourage GID diagnosis and treatment. And Scott does become the brunt of homophobic harassment when he arrives at school with his fingernail painted red. The decision to dye his hair and to paint one of his fingers and toes with red nail polish—which he defines as "girlish sort of things"—appears to be an act of working at the limits of a particular model of heterosexual masculinity that is imposed upon him, on this occasion, by his parents (see Foucault, 1987, 1984b). In some sense, the fact that he was not accepted by many of the "footballers" at school and was continually harassed and called names, with his sexuality being constantly brought into question, may also account for his desire to engage in transgressive self-fashioning practices. His behavior may be read as a form of protest against the regulatory norms of masculinity governing the behavior and conduct of the "footballers."

Likewise, in choosing to do weightlifting during adolescence, a sport traditionally considered the domain of men with its emphasis on musculature, Saree Williams (1998:163) found her sexuality being interrogated, and she had to "educate" herself into adopting resistant positions: "to learn not to listen to people who said things like 'Weightlifter? You must be gay.' What has my sexual orientation got to do with me competing in a sport or not?"

Stephen, self-labeled as a "tomgirl," talks about the dilemmas heteronormative and gender-normative structures and practices construct for young people wishing to explore sexual diversity beyond sexual duality. At the time of the interview, Stephen was hoping to find a girlfriend to "find out if I'm gay."

> Maria: You know how you talked about having a girlfriend to see if you're gay; have you ever thought of doing it the other way, having a boyfriend to see if you're straight or gay?
> Stephen: That might be hard because there's not many gay kids; there's no gay kids in this school. I don't think so. It would be pretty hard, I'll probably have to have a girlfriend...Actually I've felt sad about that but, yeah, I thought I might be bisexual.
> Maria: What does that mean to you?
> Stephen: Where guys like girls and boys, and girls like girls and boys, yes, bisexual.
> Maria: And would that worry you if that's what you were?
> Stephen: It would a bit, but I wouldn't mind it; it would be a bit hard to cope with.
> Maria: Why do you reckon it might be hard to cope with?
> Stephen: I want to be myself like every other kid, be normal and not like both guys and girls.
> Maria: So do you reckon it's more normal just to like guys or girls?
> Stephen: It's normal being gay too, but it's just really hard, it's going to be really hard if I am.

Here we see an example of the use of panopticonic silence that prevents any other option for Stephen to explore in order to situate him/herself as gay, heterosexual or bisexual. Options and strategies are severely limited within the heteronormative context of the school and the homophobia of other male students. Indeed, Stephen seems aware of a socially constructed hierarchy, with heterosexuality being deemed the most normal to bisexuality being the most abnormal.

Beyond Duality: Silencing Transgender Identities in Schools

> When I was in high school everybody thought I was a boy—they even called me Larry. This wasn't surprising though, because I was born with a penis and raised as a boy. But I knew I wasn't—I knew secretly that I was a girl. I kept this to myself because the secret seemed so terrible and I thought that if I told someone they'd think I was mad. Things were so confusing for me at school, I didn't know what I was, and was afraid to find out. (Seabrook, 1998:226)

Transgenderism and intersexuality (hermaphroditism) stand as glaring testimonies to the fact that the line between the sexes is not as firm and rigid as binarily constructed theories of sex and gender propose. To date, Australian schools have not taken up this issue. Despite the increasing sociocultural and political acknowledgment and activism in relation to gender diversity and sexual diversity, current debates, policies, and programs are still framed by notions of gender duality. Masculinities and femininities in various sociocultural forms are increasingly becoming acceptable or at least debated, as this chapter has illustrated, but strictly in two oppositional and fixed bio-logical forms (Pallotta-Chiarolli, 1999b). When Stephen was asked to define a "tomgirl's" gender during the interview, the response could easily be construed as a case of GID. An imaginary line with boys at one end and girls at the other was drawn in the air and Stephen indicated "right in the middle."

> Maria: And why do you think that's so?
> Stephen: Because I'm different, I'm both, I dress up as a girl sometimes and a boy sometimes.

Stephen's situation is an example of the possibilities that can be explored beyond diagnosis and treatment for GID. Stephen had experienced severe harassment at his/her primary school before his/her parents found a secondary school that was committed to accepting Stephen's cross-gender/transgender explorations and minimizing harassment. Stephen's parents, particularly his/her mother, did express concern about Stephen's cross-dressing and behavior, and his welfare at school and in society. Nevertheless, they were collaborating closely with the school principal and teachers to focus on Stephen's emotional well-being and educational achievements while the young adolescent was given some latitude to explore various gender and sexual roles. This involved educating and monitoring other students in relation to their understandings of gender, sexuality, and harassment of Stephen.

This has not always been the situation for older transgendered people. In an interview, Jacqui Cussen, a transgender woman, recalls how "a massive fear took over" in secondary school as she realized that although she believed she was a girl, everyone else defined her as a boy and treated her accordingly: "I thought that if anybody knew, then I'd lose everybody. I'd be totally alone, and totally isolated, and then I created the barriers and isolated myself."

Here we see the direct interconnection of panopticonic surveillance and self-regulation. It was only after she left the school where there had been absolute silence on these issues that Jacqui discovered "there were other people," only to discover the world of regulated performance of transgenderism:

> Jacqui: Les Girls, being a showgirl, which was not for me as well.
> Maria: Can you explain how you felt?
> Jacqui: I'd finally found some people who seemed to be in many ways like me, born as boys and living as women, but the way they were living was outrageous, that wasn't me, but if you were transgendered, that was what you have to be. I had to become something else that's not me.

Intersexual persons are designated as "freaks" in Western cultures and are encouraged to normalize by assuming either a male or female body, thereby having a gender imposed upon them:

> Instead of allowing a third, fourth, and even fifth sex to develop (given the three types of hermaphrodites), we engage in surgery before the age of consent to eliminate each of these alternative genders to fix an individual purely in the male or female category. Thus, hermaphrodites become an invisible aspect of the male and female poles. (Colker, 1996:89–91)

The ambiguity and incongruity of the "it" that does not pass as a person of a biological singular sex cannot be tolerated. It must become a more acceptable "him" or "her" even if medicalized normalization is "abnormalization" for some intersexual and transgendered persons. The sex classification of human beings is violated by this *sex normativity* which is then interconnected with *gender normativity* and *heteronormativity* to mean that adolescent transgendered and intersexed persons are imbricated in three systems of surveillance and self-regulation. Again, this highlights Mauss's (1973) point that the formation of human attributes cannot be confined to social, biological or psychological functions but is the effect of an ensemble of practices in which all three elements are mixed and influence the others.

The incorporation of the experiences and situations of transgendered and intersexed persons into educational debates in relation to gender and sexuality involves moving from a structure of complementarity or symmetry (male/female and heterosexual/ homosexual) to multiplicity and contextualization, in which what once stood as an exclusive dual relation between male and female becomes an element in a larger framework of multiple social relations (Pallotta-Chiarolli, 1999a). It reconfigures the relationships between the original pair, and puts into question identities previously conceived as stable, unchallengeable, grounded, and "known" (Garber, 1992; Pratt, 1995). Likewise, the whole issue of the coercion to cross over to the other gender via medical procedures, if one is not to remain as one is, needs to be critiqued and broadened. For example, from a postcolonial anthropological perspective that has been re-claiming and exploring precolonial gender diversity, Nanda writes,

> by defining the transsexual as a *transitional* status,...a person is a transsexual when he or she is in a *temporary* transitional, in-between sex status...it cannot be, in our culture, a permanent possibility. The reconstructive genital surgery that transsexuals desire aims at moving them from just such an in-between state...to the status of a real woman, with female sex organs....The central theme of transsexualism as a transitional stage is "passing," that is, learning how to conceal the fact that one is a transsexual. (Nanda, 1990:137–138)

Hence, as in precolonial American Indian cultures, whereas Westerners feel uncomfortable with the ambiguities and contradictions inherent in such in-between categories as hermaphroditism and transgenderism and playing with gender by young people like Stephen, Hinduism not only accommodates such ambiguities, but also views them as meaningful and even powerful (Nanda, 1993; Kumar, 1993). In Thailand, the term *kathoey* (transvestite/transsexual/ hermaphrodite) means neither male nor female, but both: "a coherent identity attached to diverse and fluid practice," a combination of fixed subjectivity and multiple subject positions (Morris, 1994:19).

A broader framework that encompasses multiple possibilities from surgical intervention to nonintervention and the multiple ways of performing gender within and between these options is provided by the self-defined "gender-ambiguous" Jillian Hooley (1997:13) who uses the term "transgender" as an umbrella label that refers "to those who live outside the norms, ideals and conventions of gender. It is intended as a status which describes fluid social practices or behaviour rather than a category, although there is a tendency toward categorical reversion, which seems difficult to avoid."

Some transgender persons are re-inscribing and reconstituting themselves outside the binary medical framework, rejecting the labels "transsexual" and "transvestite" as connoting sickness and emotional deficiency. These labels also "mandated the taking-on of membership, and accompanying regulatory practices and disciplines of the body required for membership." Thus, according to Hooley (1997: 32), the old "trapped in the wrong body story" is an example of "tired, essentialist narratives" that construct gender only in dichotomous biological terms.

It appears that if transgendered persons, and the wider society they inhabit, may come to see identity as "shifting, constructed, contingent, relational—or as fluid rather than fixed," they may live their bodies rather more creatively, healthily and cheaply than the "technological fix" which transsexual medicine provides. Ambiguity may be regarded "as a strength rather than symptomatic of a 'sick,' indecisive, or confused identity" (Hooley, 1997:33). Hence, gender reassignment as a result of GID is seen as potentially foreclosing "the possibility of a life grounded in the *inter-textual* possibilities of the transsexual body" (Hooley, 1997:33–34). However, this queer or multiple gender, or life on the borderlines of sexual identity, may not appeal to all in the "transgender community." Transgender activists as well as feminists need to be accepting of intra-diversity and not be too regimented and regulatory in relation to those who do wish to become the other, so-called opposite, gender. Again, the framework needs to be shifted from duality to multiplicity so that transgendered, cross-gendered and gender explorative young people are able to consider possibilities for themselves as gendered and sexual beings from a wide range of options including surgical intervention, hormonal intervention, cross-dressing and other body-fashioning practices, and nonintervention of any kind (Pallotta-Chiarolli, 1996, 1999a).

What is required in schools is a greater engagement with these issues for all adolescents within the curriculum and the provision of an environment where young people such as Stephen can experiment and explore and in doing so educate their self as well as educate others around them. All students are negotiating the performance of gender, whatever their sexuality. There needs to be an opening up of possibilities that there is a third position or multiple positions outside the duality of male and female sex and gender that imply the self-fashioning of identity from a broader array of available biological and sociocultural alternatives (Pallotta-Chiarolli, 1996, 1999a). As Jacqui explains in her interview,

> unless the norms of two fixed genders are shifted, then being anything but those two appropriate genders is called things like cross-dressing, transgender, which always mean transgression of a norm, a negativity, and requiring people like me to explain and justify ourselves and our positions.

Conclusion: Beyond Gender and Sexual Normativity

As a way of moving beyond the binary of gender duality and transgenderism, Jacqui Cussen proposes the use of the word "orlando" based on Virginia Woolf's (1920) character in her book, *Orlando*, who was born as one sex and gradually developed into another, while maintaining elements of both.

> A word that admits all those stages of myself without focusing on one or more aspect of myself at the expense of the others....I am an orlando. I have an orlandic nature. (Cussen, 1998:3–4)

Sharon Minter (1999:28) believes we can learn much from work that has already been done in relation to gay and lesbian sexualities:

> As researchers have examined lesbian, gay, and bisexual youth without the preconception of inherent pathology, they have found (1) that sexual minority youth who have adequate social support are no more likely to experience mental health problems than their heterosexual peers; (2) that sexual minority youth who do not have adequate support are at heightened risk of depression and other mental heath problems as a result of isolation and/or rejection from family and peers; and (3) that the most effective interventions for these youth are those that focus on alleviating social isolation and other external stressors, including referrals to support groups and other community resources.

Minter (1999:29), distinguishes between

> defining gender atypicality in children as a disorder and identifying gender-variant children as a distinct group who may be at heightened risk for depression and other mental health problems because of social stigma and rejection.

Therefore, therapists need to "redirect the energies and research monies that continue to be spent on the dubious and futile quest to eliminate gender atypicality in children" to focus on the far more rational, ethical and achievable goal of how to help gender-variant children develop the emotional and social resources needed by every child who is at risk for stigma and isolation. In the meantime, there is little doubt that current clinical approaches to GID in children reinforce the larger society's rejection and disapproval of gender-variant children, or that current treatments have damaged children who already confront formidable obstacles to developing and maintaining self-esteem. (29)

Research into gender and sexuality within the context of schooling needs to pay greater heed to the experiences and situations of gender-variant, transgendered, and intersexed persons who are currently hidden or ignored by educational researchers as well as ignored within schools and concealing themselves out of fear of violent and oppressive repercussions (Pallotta-Chiarolli, 1999b). As Lingard and Douglas (1999:127) write, "Schools need to be involved in the processes of constructing what Connell (1995) has called 'gender multiculturalism'—a postmodern multiplicity of acceptable ways of performing gender."

The ultimate effect of such an approach will be to reconfigure the rigid heteronormative boundaries governing male/female and heterosexual/homosexual binary categorizations, which are institutionalized within the social and educational contexts from which the subjects of this research speak and write. In this way ambiguity and the in-between categories can be embraced as a strength rather than as symptoms of a deviant or confused identity (Pallotta-Chiarolli, 1999a).

In this chapter we have attempted to spell out the effects of such sex-normative, gender-normative and heteronormative practices and modes of relating in schools and the implications for further developing a theorization of gender and sexuality that embraces what Cussen (1998) terms *orlandic* principles and modes of thinking. We believe that the experiences and insights young people offer about the policing of gender and sexuality in their lives signal the potentially disruptive work that can be achieved in schools through involving students in active discussions about the effects of sex/gender systems that naturalize compulsory heterosexuality, the effect of which is to relegate gender and sexual ambiguity to the realm of pathological deviancy (Martino and Pallotta-Chiarolli, 2001). We see our research as a gesturing toward a political realization of Cussen's *orlan* possibilities in creating spaces for educational systems to acknowledge the multiplicity of gender performances alongside a developing awareness of cultural and sexual diversity. The voices of the students included in this chapter attest to the exciting possibilities that exist for opening up such discussions in schools.

> As we continue to recognize the complexities of sexualities and gender norms we become even more aware of the harmful consequences of dichotomizing and emphasizing adherence to strict gender role stereotypes. We need to embrace, rather than pathologise, label as deviant, and then try to change these so called "gender disturbed children." For they have much to teach us about accepting our own diversity. (Neisen, 1992:67)

Acknowledgments

We acknowledge with gratitude the work of Jacqui Cussen who provided us with a new definition and fresh insight. Also, a thank you to Greg Curran, Michael Crowhurst, and Lyn Scott in assisting us with contacting participants in our research. We also acknowledge the assistance of Dr. Felicity Haynes in finding some valuable material. And finally to the young people such as Stephen with whom we have worked and who have privileged us with such personal insights into their lives, we thank you and hope you feel that your contributions have been put to good use. We are humbled by your pioneering insights and strengths in living out your gender and sexual realities despite enormous pressure to do otherwise. This chapter is for you.

References

Anzaldua, Gloria (ed.). (1990). *Making faces, making soul/Haciendo caras.* San Francisco: Aunt Lute.

Butler, Judith. (1990). *Gender trouble: Feminism and the subversion of identity.* New York: Routledge.

Butler, Judith. (1991). Imitation and gender insubordination. In D. Fuss (ed.), *Inside/out: Lesbian theories, gay theories.* New York: Routledge.

Butler, Judith. (1993). *Bodies that matter: On the discursive limits of "sex."* New York: Routledge.

Butler, Judith. (1996). The poof paradox: Homo-negativity and silencing in three Hobart high schools. In L. Laskey and C. Beavis (eds.), *Schooling and sexualities: Teaching for a positive sexuality.* Geelong: Deakin University for Education and Change.

Coates, Susan. (1987). Extreme femininity in boys. *Medical Aspects of Human Sexuality*, August, 104–110.

Cohen, Ira J. (1987). Structuration theory and social praxis. In A. Giddens and J. H. Turner (eds.), *Social theory today.* Oxford: Polity Press.

Cohen, Stanley, and Taylor, Laurie. (1976). *Escape attempts: The theory and practice of resistance to everyday life.* London: Allen Lane.

Colker, Ruth. (1996). *Hybrid: Bisexuals, multiracials and other misfits under American law.* New York: New York University Press.

Connell, R. W. (1987). *Gender and power: Society, the person and sexual politics.* Cambridge: Polity Press.

Connell, R. W. (1992). A very straight gay: Masculinity, homosexual experience, and the dynamics of gender. *American Sociological Review*, Vol. 57, 735–751.

Connell, R. W. (1995). *Masculinities*. Sydney: Allen & Unwin.
Cussen, Jacqui Zephyr. (1998). A rose by any name. Unpublished paper cited by courtesy of the author.
Dancer, Tamsin. (1998). Take three steps back (into the closet). In M. Pallotta-Chiarolli (ed.), *Girls talk: Young women speak their hearts and minds*. Lane Cove, Sydney: Finch Publishing.
Dixon, C. (1997). Pete's tool: Identity and sex-play in the Design and Technology classroom. *Gender and Education*, Vol. 9, No. 1, 89–104.
Epstein, D. (1994). *Challenging lesbian and gay inequalities in education*. Buckingham, England, and Philadelphia: Open University Press.
Epstein, D. (1997). Boyz' own stories: Masculinities and sexualities in schools. *Gender and Education*, Vol. 9, No. 1, 105–115.
Foucault, Michel. (1977). *Discipline and punish: The birth of the prison*. New York: Vintage Books.
Foucault, M. (1980). *Michel Foucault: Power/knowledge: Selected interviews and other writings 1972–1977*. C. Gordon (ed.). Hassocks, England: Harvester.
Foucault, M. (1984a). What is enlightenment? Trans. Catherine Porter. In P. Rabinow (ed.), *The Foucault Reader*. London: Penguin.
Foucault, M. (1984b). Preface to *The history of sexuality*, Volume II. In P. Rabinow (ed.), *The Foucault Reader*. London: Penguin.
Foucault, M. (1985). *The history of sexuality*: Volume 2. Trans. R. Hurley. New York: Vintage.
Foucault, M. (1986). *The history of sexuality:* Volume 3. Trans. R. Hurley. New York: Vintage.
Foucault, M. (1987). The ethic of care for the self as a practice of freedom. *Philosophy and Social Criticism*, Vol. 12, 113–131.
Foucault, M. (1991). Politics and the study of discourse. In G. Burchell, C. Gordon and P. Miller (eds.), *The Foucault effect: Studies in governmentality*. London: Harvester Wheatsheaf.
Frank, B. (1993). Straight/strait jackets for masculinity: Educating for real men. *Atlantis*, Vol. 18, Nos. 1 & 2, 47–59.
Garber, Marjorie. (1992). *Vested interests: Cross-dressing and cultural anxiety*. London: Penguin.
Haywood, C. (1993). Using sexuality: An exploration into the fixing of sexuality to make male identities in a mixed sexed sixth form. M.A. (Sociology of Education) Dissertation, University of Warwick.
Hirst, P., and P. Woolley. (1982). *Social relations and human attributes*. London: Tavistock.
Hite, S. (1981). *The Hite Report on male sexuality*. New York: Alfred A. Knopf.
Holland, J., C. Ramazangolu, and S. Sharpe. (1993). *Wimp or gladiator: Contradictions in acquiring masculine sexuality*. London: Tufnell Press.
Holmes, Morgan. (1994). Re-membering a queer body, *Undercurrents*, May, 11–13.
Hooley, Jillian. (1997). Transgender politics, medicine and representation: Off our backs, off our bodies. *Social Alternatives*, Vol. 16, No 1, 31–34.
Hunter, I. (1991). From discourse to dispositif: Foucault and the study of literature. *Meridian*, Vol. 10, 36–53.

Jordan, E. (1995). Fighting boys and fantasy play: The construction of masculinity in the early years of school. *Gender and Education*, Vol. 7, No. 1, 69–86.

Kazmi, Yedullah. (1993). Panopticon: A world order through education or education's encounter with the Other/difference. *Philosophy and Social Criticism*, Vol. 19, No. 2, 195–213.

Kehily, M., and A. Nayak. (1997). Lads and laughter: Humour and the production of heterosexual hierarchies. *Gender and Education*, Vol. 9, No. 1, 69–87.

Kessler, S., D. Ashenden, R. Connell, and G. Dowsett. (1985). Gender relations in secondary schooling. *Sociology of Education*, Vol. 58, 34-88.

Kessler, Suzanne J. (1998). *Lessons from the intersexed*. New Brunswick, NJ: Rutgers University Press.

Kumar, Arvind. (1993). Hijras: Challenging gender dichotomies. In R. Ratti (ed.), *A lotus of another color: An unfolding of the South Asian gay and lesbian experience*. Boston: Alyson Publications.

Langley, Jess. (1998). Coming out/going home. In M. Pallotta-Chiarolli (ed.),*Girls talk: Young women speak their hearts and minds*. Lane Cove, Sydney: Finch Publishing.

Laskey, L., and C. Beavis (eds.). (1996). *Schooling and sexualities: Teaching for a positive sexuality*. Geelong: Deakin University Centre for Education and Change.

Lingard, Bob, and Peter Douglas. (1999). *Men engaging feminisms: Pro-feminism, backlashes and schooling*. Buckingham, England: Open University Press.

Mac an Ghaill, M. (1994). *The making of men; masculinities, sexualities and schooling*. Buckingham, England, and Philadelphia: Open University Press.

Martino, Wayne. (1998a). Girls talk school, boys and friendships In M. Pallotta-Chiarolli (ed.), *Girls talk: Young women speak their hearts and minds*. Lane Cove, Sydney: Finch Publishing.

Martino, Wayne. (1998b). Interrogating masculinities: Regimes of practice. Doctoral thesis, School of Education, Murdoch University.

Martino, Wayne. (1998c). When you only have girls as friends, you've got some serious problems: Interrogating masculinity and homophobia in the critical literacy classroom. In M. Knobel, and A. Healy. (eds.), *Critical literacies in the primary classroom*. Sydney: PETA.

Martino, Wayne, and Maria Pallotta-Chiarolli. (forthcoming, 2001). *So what's a boy? Addressing issues of masculinity and schooling*. London: Open University Press.

Mason, G., and S. Tomsen (1997). *Homophobic violence*. Leichardt, NSW: Federation Press.

Mauss, M. (1973). Techniques of the body. Trans. B. Brewster. *Economy and Society*, Vol. 2, 70–87.

mAy-welby, norrie. (1998). Having both genders. In M. Pallotta-Chiarolli (ed.), *Girls talk: Young women speak their hearts and minds*. Lane Cove, Sydney: Finch Publishing.

Miller, Bethwyn. (1998). Untitled. In M. Pallotta-Chiarolli (ed.), *Girls talk: Young women speak their hearts and minds*. Lane Cove, Sydney: Finch Publishing.

Minter, Sharon. (1999). Diagnosis and treatment of Gender Identity Disorder in children. In Matthew Rottnek (ed.), *Sissies and tomboys: Gender nonconformity*

and homosexual childhood. New York: New York University Press.
Morris, Rosalind C. (1994). Three sexes and four sexualities: Redressing the discourses on gender and sexuality in contemporary Thailand. *Positions*, Vol. 2, No. 1, 15–43.
Nanda, Serena. (1990). *Neither man nor woman: The Hijras of India.* Belmont, CA: Wadsworth Publishing.
Nanda, Serena. (1993). Hijras as neither man nor woman. In H. Abelove, M. A. Barale, and D. M. Halperin (eds.), *The lesbian and gay studies reader.* New York: Routledge.
Nayak, A., and M. Kehily. (1996). Playing it straight: Masculinities, homophobias and schooling. *Journal of Gender Studies*, Vol. 5., No. 2, 211–230.
Neisen, Joseph. (1992). Gender identity disorder of childhood: By whose standard and for what purpose? A response to Rekers and Morey. *Journal of Psychology and Human Sexuality*, 5(3): 65–67.
Pallotta-Chiarolli, Maria. (1996). Only your labels split the confusion: Of impurity and unclassifiability, *Critical Inqueeries*, Vol. 1, No. 2, 97–118.
Pallotta-Chiarolli, Maria (ed.). (1998). *Girls talk: Young women speak their hearts and minds.* Lane Cove, Sydney: Finch Publishing.
Pallotta-Chiarolli, Maria. (1999a). Mestizaje tapestry: Interweaving multi-culturalism, multi-sexuality and multi-partnering. Doctoral Thesis. Deakin University, Geelong.
Pallotta-Chiarolli, Maria. (1999b). Review of *Masculinity goes to school* (P. Gilbert and R. Gilbert) and *Schooling sexualities* (D. Epstein and R. Johnson), *Australian Educational Researcher* Vol. 26, No. 1, 98–103.
Parker, A. (1996). The construction of masculinity within boys' physical education. *Gender and Education*, Vol. 8, No. 2, 114–157.
Pendrey, Catherine. (1998). Untitled. In M. Pallotta-Chiarolli (ed.), *Girls talk: Young women speak their hearts and minds.* Lane Cove, Sydney: Finch Publishing.
Plummer, Kenneth. (1981). *The making of the modern homosexual.* London: Hutchinson.
Pratt, Minnie Bruce. (1995). *S/he.* Ithaca, NY: Firebrand Books.
Pursey, Belinda. (1998). Evolutionary love. In M. Pallotta-Chiarolli (ed *Girls talk: Young women speak their hearts and minds.* Lane Cove, Sydney: Finch Publishing.
Rabinow, Paul. (1984). *The Foucault reader.* London: Penguin.
Rajchman, J. (1986). Ethics after Foucault. *SocialText*, Vol. 13, 165–183.
Redman, P. (1996). Curtis loves Ranjit: Heterosexual masculinities, schooling and pupils' sexual cultures. *Educational Review*, Vol. 48, No. 2, 175–182.
Rogers, Michelle. (1998). Being on patrol and being patrolled. In M. Pallotta-Chiarolli (ed.), *Girls talk: Young women speak their hearts and minds.* Lane Cove, Sydney: Finch Publishing.
Rottnek, Matthew (ed.). (1999). *Sissies and tomboys: Gender nonconformity and homosexual childhood.* New York: New York University Press.
Sally, Lacey, and Grandma. (1998). Your Mum's a lezzo. In M. Pallotta-Chiarolli (ed.), *Girls talk: Young women speak their hearts and minds.* Lane Cove,

Sydney: Finch Publishing.
Seabrook, Laura Anne (1998). When your gender doesn't feel right. In M. Pallotta-Chiarolli (ed.), *Girls talk: Young women speak their hearts and minds*. Lane Cove, Sydney: Finch Publishing.
Signorile, M. (1997). *On the outside*. New York: Verso.
Skelton, C. (1997). Primary boys and hegemonic masculinities. *British Journal of the Sociology of Education*, Vol. 18, No. 3, 349–369.
Smith, Heather. (1998). Following the gender rules. In M. Pallotta-Chiarolli (ed.), *Girls talk: Young women speak their hearts and minds*. Lane Cove, Sydney: Finch Publishing.
Steinberg, D. L., D. Epstein, and R. Johnson. (1997). *Border patrols: Policing the boundaries of heterosexuality*. London: Cassell.
Walker, J. (1988). *Louts and legends: Male youth culture in an inner-city school*. Sydney, London, and Boston: Allen & Unwin.
Ward, N. (1995). 'Pooftah,' 'wanker,' 'girl': Homophobic harassment and violence in schools. In *Girls and boys: Challenging perspectives, building partnerships* (Proceedings of the Third Conference of the Ministerial Advisory Committee on Gender Equity). Brisbane: Ministerial Advisory Committee on Gender Equity.
Williams, Saree. (1998). Weightlifter? You must be gay. In M. Pallotta-Chiarolli (ed.), *Girls talk: Young women speak their hearts and minds*. Lane Cove, Sydney: Finch Publishing.
Willis, P. (1977). *Learning to labour: How working class kids get working class Jobs*. Westmead, England: Saxon House.
Wood, J. (1984). Groping towards sexism: Boys' sex talk. In A. McRobbie and M. Nava (eds.), *Gender and generation*. London: Macmillan.
Woolf, Virginia. (1920). *Orlando*. London: Hogarth Press.
Zucker, Kenneth J., and Susan J. Bradley. (1995). *Gender Identity Disorder and psychosexual problems in children and adolescents*. New York: Guilford Press.

2. WAYS OF UNDERSTANDING OTHERS

> Man's quest for certainty is, in the last analysis, a quest for meaning. But the meaning lies buried within himself rather than in the void he has vainly searched for portents since antiquity.
> —Loren Eiseley, *The Man Who Saw through Time*

The articulation of a generic social process of "male femaling" or genderblending elaborated by Richard Ekins provides a conceptual framework for a fairly orthodox sociology of transgendering stories. Following Plummer's work on sexual stories, Ekins and Dave King explore contemporary transgender diversity by grouping their transgendering stories into four major modes or styles which they term "migrating," "oscillating," "negating," and "transcending." They give illustrative examples of each mode with reference to the binary male/female divide, the interrelations between sex, sexuality, and gender, and the interrelations between the four main sub-processes of transgendering which they identify as "substituting," "concealing," "implying," and "redefining." Sexuality is one other aspect of gender that is either concealed or dehumanized in institutions.

Johnson moves away from a focus on the cause of transsexualism, as theorized by what she calls the medical model, toward the study of transsexual identity. She analyzes two methods, the participant observation utilized by Bolin, and the grounded theory employed by Ekins, before presenting the merits of a discursive approach. She argues that discourse analysis permits notions of subjectivity, self, and identity to be explored in a way that enables us to address the complexity of a transsexual identity. A greater understanding of the experience of being transsexual can only be of benefit if we are to see transsexuals accepted as valid and valued members of society. As with Ekins' and King's stories, the meaning of personal identity is revealed to have more to do with interpersonal communication and practices than with a scientific theory.

Surya Monro shows that although feminists rendered patriarchal dominance theoretically empty by exposing its assumptions they are still dominated in practice by those assumptions. Similarly, when modern technology renders biological sex mutable, politics based on binaries dissolves only theoretically. Struggles in gender politics rage between feminists, queer and transsexual/transgender activists and within transsexual/transgender communities, demonstrating the persistence of "real" conceptual barriers provided by a deeply transphobic and hetero-patriarchal society. "Gender freedom" calls for a practical recognition of diversity, agency, non-harm, and equality.

Tales of the Unexpected: Exploring Transgender Diversity through Personal Narrative

Richard Ekins and Dave King

> In the dressing room, I quickly tried the skirt on. It fit as though it were made for me. I turned around in front of the large mirror in the showroom. My aunt was enthusiastic. "We're buying it. Leave it on."

Out of context there is nothing unusual about the incident reported here. In context, however, it becomes part of one of many "Tales of the Unexpected," for the narrator is a boy and the aunt is cross-dressed as a man (von Mahlsdorf (1995:51–52). During World War II, von Mahlsdorf murdered his father, called himself Charlotte (after his cross-dressing lesbian aunt's lover), and has lived openly as a transvestite ever since. Charlotte, well past 60 when his autobiography was first published, is "a quietly passionate, steadfast and serene figure" (according to the book-cover blurb) who shuns make-up, wears the most simple frocks and has become both "his own woman" and "his own wife."

In order to provide a framework for the analysis of such tales of the unexpected that pays the proper respect to contemporary transgender diversity, this chapter develops the conceptual framework introduced in Ekins and King (1999, 2000). In this recent work, we emphasized transgendered bodies (sex) and "selling body transgendering stories." Here we broaden our scope to cover issues of sexuality (the erotic) and gender (the social and cultural correlates of sex), relating to a wider range of transgendering personal narratives.

Gendering and Transgendering

For sociologists, gender is a system of social differentiation and social placement. Societies have understandings about what constitutes gender, how many gender categories there are, who belongs to which category, what characterizes members of each category and so on. This is what Ramet (1996) refers to as a "gender culture." A society's understanding of gender is expressed in terms of a complex set of rules that tell us what to expect of other people's behavior in both a predictive sense (what will happen) and in a normative sense (what should happen).

Gender is, of course, an important part of individual identity, but following Garfinkel (1967) and Kessler and McKenna (1978), we find it useful not to think of gender as something that people have, but to see the production of a gendered social identity as an ongoing accomplishment; something that is constantly being "done." So we use the verb "gendering," here, to refer to the processes whereby a person is constituted as gendered on an everyday basis. In a culture such as ours which recognizes only two genders, gendering can be divided into two processes, "maling" and "femaling." A basic rule of our gender culture is that only biological males are expected to "male," and only biological females are expected to "female." Where this rule is broken—where males "female" (Ekins, 1993, 1997) and females "male" (Ekins, 1984)—we use the term transgendering (Ekins and King, 1996b, 2000). Transgendering stories are, therefore, "Tales of the Unexpected." In one way or another, they breach the "rules" of gender.

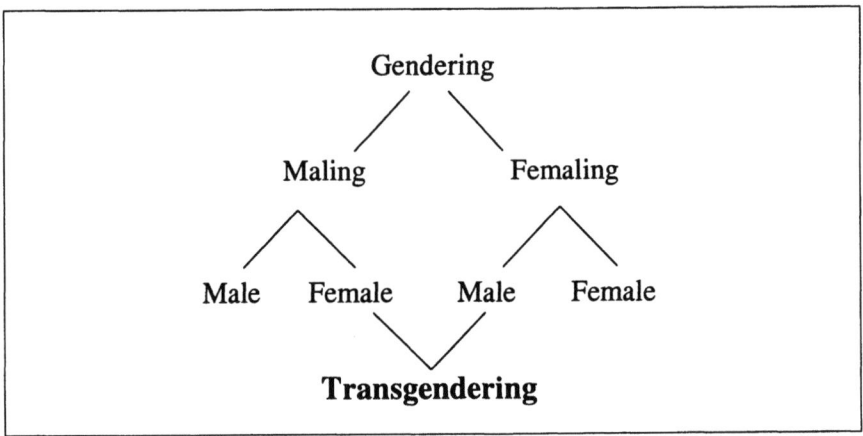

Figure 2. Transgendering

Modes and Processes of Transgendering

Stories come in many forms but here our concern is with the personal narrative (Plummer, 1995, 1996). A number of conceptually distinct contemporary transgendering stories have emerged from our two decades of qualitative sociological research with several thousand transpeople in Western Europe, North America, South Africa, Australia, and New Zealand (e.g., Ekins, 1983, 1993, 1997; King, 1981, 1993; Ekins and King, 1996a). These stories have meaning only in relation to their cultural context—that is, they only make sense as transgendering stories if we situate them in relation to what is assumed about gendering. The dominant Western understanding of gender is that it is a binary system given by "nature." In the main part of this chapter we discuss transgendering stories in terms of four main modes depending on their relationship to the male/female binary divide. We have termed the four modes "migrating," "oscillating," "negating," and "transcending." "Migrating" denotes a mode of transgendering that involves moving from one side of the binary divide to the other on a permanent basis. "Oscillating" entails moving backward and forward over the gender border, only temporarily resting on one side or the other. "Negating" denotes the "ungendering" of those who seek to nullify maleness/masculinity or femaleness/femininity and deny for themselves the existence of a binary divide. "Transcending" involves going "beyond gender," into a third space.

What ethnomethodologists refer to as the "natural attitude" toward gender (Garfinkel, 1967:122–128; Kessler and McKenna, 1978:113–114) assumes that all human beings will belong to one of two discrete gender categories permanently determined on the basis of biologically ("naturally") given characteristics. This "natural attitude" specifies that everybody must be either male or female. A person cannot be both male and female or neither. Not only are male and female held to be discrete categories, they are also held to be opposites.

Given the above, transgendering is a problematic process. Our data suggests five main sub-processes by which transgendering, in any one of the four modes, is accomplished. The first sub-process involves "erasing," which entails the eliminating of aspects of maleness or femaleness, masculinity or femininity. A genetic male may undergo castration. A genetic female may undergo a hysterectomy. Both males and females may wear unisex clothes and adopt ungendered mannerisms.

The second sub-process involves "substituting." The person who is transgendering may (with or without outside help) replace the body parts, identity, dress, posture, gesture, and speech style, that are associated with one gender, with those associated with the other. The degree of substitution will depend on a number of factors such as the particular personal project of the individual, the personal circumstances, the development of any technology and aids that may be used, and (where not covered by some healthcare scheme) the financial resources to afford them. Substituting may be temporary or permanent.

"Concealing," as the third sub-process, refers to the concealing or hiding of things that are seen to conflict with the intended gender display. It may, for instance, involve hiding body parts—wrapping a scarf around the Adam's apple, tucking the penis, binding the breasts, and so on. It may involve the concealing of details of biography that are gender-specific—a marriage, a birth certificate, and so on. For female malers, displaying male characteristics may be more important than concealing female ones if Kessler and McKenna (1978: 158–159) are correct when they argue that: "In order for a female gender attribution to be made, there must be an absence of anything which can be construed as a 'male only' characteristic. In order for a male gender attribution to be made, the presence of at least one 'male' sign must be noticed, and one sign may be enough." The basic categorizing "schema" is "See someone as female only when you cannot see them as male" (Kessler and McKenna, 1978:158).

In addition to concealing and displaying, transgendering may involve "implying" certain body parts or gendered attributes. Because the body is usually apprehended in social interaction in its clothed form, it is possible to imply the gendered form of the body beneath. So, for example, males can wear breast forms inside a bra, or hip pads inside a panty girdle; females may place something in their underpants to imply the possession of a penis. Implying may be the only sub-process involved in "virtual gender swapping" on internet discussion lists or, indeed, in any situation where interaction is not face to face, such as situations involving the telephone or written communication.

The fifth sub-process is "redefining." Whereas the meanings of erasing, substituting, concealing, and implying are relatively easily grasped, particularly in relation to the acceptance of the binary divide, redefining is more subtle and multilayered. At one level, the nature of the body, body parts, and gendered accompaniments may be redefined. The male to female transsexual may redefine her beard growth as facial hair. The penis may be redefined as a "growth between the legs," as in "I was a woman who had needed some corrective surgery. The growth was gone and my labia, clitoris and vagina were free" (Spry, 1997:152). The male transvestite may redefine unisex clothes, such as jeans and T-shirts, as women's clothes because they were bought from a women's boutique. Moreover, in the case of transcending—which seeks to subvert and/or move beyond the binary divide—the process of redefining the binary divide may entail selves, bodies, body parts, and gendered accoutrements taking on new meanings within the redefined system of classification.

Migrating Stories

The *Concise Oxford Dictionary* (1995) defines migrating as "moving from one place of abode to another especially in a different country." Many gender migrants speak of starting a new life in a new country or of being reborn. While return may be possible, at its inception the journey is seen as one-way; it is not expected that there will be any turning back.

The male to female transsexual, McCloskey (1999:xi–xii), highlights the essence of migrating:

> But people do after all cross various boundaries. I've been a foreigner a little, in England and Holland and on shorter visits elsewhere. If you've been a foreigner you can understand somewhat, because gender crossing is a good deal like foreign travel. Most people would like to go to Venice on vacation. Most people, if they could magically do it, would like to try out the other gender for a day or a week or a month...But only a tiny fraction of the crossgendered are permanent gender crossers, wanting to become Venetians. Most people are content to stay mainly at home. A tiny minority are not. They want to cross and stay.

Migrating stories are part of a broader category of "modernist tales." Plummer (1995:49–50) outlines the main elements of one major type of modernist story as beginning with pain and suffering, usually in silence and secrecy. Then there is a crucial turning point, a new understanding and action, which leads to a transformation a triumph, victory over suffering. Typical plots (forms) involve journeys, searches or quests and the establishment of a home (finding oneself). Some of these elements are clearly evident in the titles of some transsexual autobiographies: *April Ashley's Odyssey* (Fallowell and Ashley, 1982) and *What Took You So Long: A Girl's Journey to Manhood* (Thompson and Sewell, 1995).

In migrating stories, the prominent sub-process of transgendering is that of substituting, which may be variously progressive, problematic, rapid, and extensive. Erasing is co-opted in the service of substituting. Complete body substitution (transformation), however, is not possible except in fantasy. Modern medical interventions can accomplish a great deal, especially if administered early in the life of the body. Even so, chromosomal and gonadal substitutions are not possible and after puberty certain aspects of the body, such as height or skeletal shape, may be beyond substitution. To varying degrees, depending on the individual's personal project, physique, and the particular setting, the transgendering sub-processes of concealing, implying, and redefining will always be in play.

Published autobiographies of migrating transsexuals have become widespread in the last couple of decades. Most usually the story is told of the female trapped in the body of a male, or, somewhat less frequently, of the male trapped in the body of a female. The migrator then enlists medical aid to acquire a body to match his or her mind. As Thompson (in Thompson and Sewell, 1995:1) puts it:

> The first time I was born, it was in a body which was other than male. By some cosmic mistake, as a budding human being I had somehow chosen the wrong body, or the wrong body had chosen me. I am a transsexual person, a man really. It took me more than thirty years to reach a stage where my body started to fit my identity as a man, but now there is no doubt about it. Here I am, well and truly the male that I have always known myself to be.

There are three major variations of migrating story depending on whether the main "motive talk" (Mills, 1940; Ekins, 1983) for migrating is that of sex (the body), sexuality (the erotic), or gender (the social and cultural correlates of sex). In body-led migrating, it is variously the distaste for the genetically given body and the envy of the opposite sex body that "drives" the migrating. Rachael Webb,[1] the British transgendered activist, for instance, reports (personal communication to RE, 2000) her memories of an intense feeling of jealousy of a girl playmate's genitals at the age of six. "I thought her body was better. I disliked my own body and I intensely desired her body." With the onset of puberty, Webb's body became increasingly distasteful to her: "I found every aspect of maleness—body hair, genitals, etc., to be totally disgusting. I wanted to remove any and every aspect of my maleness."

In this story, the wish for the body of the opposite sex is not an erotic experience. Rachael wants to be desirable to men, but with the body of a woman. Referring to a time before her reassignment surgery, she relates:

> I was thinking about an older gay friend who expressed attraction and the need for affection towards me. I remember feeling confused then, and then it occurred to me, as if a revelation: of course, that's it, I want Ken to make love to me but not with me as a man. I want to have a woman's body and him to make love to me.

At the age of 40, Rachael finally embarked upon her migrating journey and, like so many who tell this story, she expresses her regret that she did not start the journey earlier. Her delay has, for instance, compromised the extent of her substituting. "I did not start taking hormones until I was 40. By that time I had, of course, developed male secondary characteristics. It will be a regret until I die that I did not start on hormones earlier."

Rachael, however, despite these misgivings, has no difficulty in passing as a woman in social settings. At times, she would like to be slimmer, younger, and better looking and would like to wear fashionable clothing. "But it's not a major part of my life." Given her lifestyle, she is happy enough to wear "slacks and blouses," remarking that "a man could wear what I wear, and no one would notice the difference" (personal communication to RE, 2000).

A much less well-known variant of the migrating story is the sexuality-led (erotic) migrating of those who are sexually aroused by the thought or image of themselves as the opposite sex. Blanchard (e.g., 1989, 1993) has developed the term autogynephilia, which is useful to identify this variant of migrating.[2] The sexual arousal may be variously to the opposite sex body, body parts, or body functions (e.g., menstruating, childbirth), or to the opposite gender display (e.g., clothes and gendered accessories). Maximum sexuality-led migrating, however, is likely to take place where body substituting is paramount and where gender and sexuality substituting are variously co-opted in the service of erotic body substituting. The U.S. activist Anne Lawrence is collecting personal narratives from transsexuals who identify with this position (1999).

One such informant is Janice. From her earliest memories, Janice had a "desperate desire" to feminize her body. She identifies with the erotic appeal of body migrating. She is now a post-operative transsexual who remains as sexually excited by her actual migrating as she was by fantasizing it:

> Nowadays my most common masturbatory fantasies usually involve little more than the sequential mental consideration of all the physical feminization I've undergone. It's like going through a list: now I have breasts; now I have a vagina; now I have hair to my shoulders; now I have pierced ears, etc. Just the contemplation of all these physical changes is enough to get me reliably excited. (personal communication to RE, 2000)

Something of the strength and power of the autogynephilic "drive" is seen in the following. Janice, at 35 years of age, has just encountered the "transvestite" and "transsexual" fantasy fiction of Sandy Thomas in a New York bookstore:

> As soon as I entered for the first time and saw the vast array of erotica, I knew I had found something special. I must have browsed for two or three hours before purchasing about a dozen books. I promptly returned with them to my hotel room and read them and masturbated to the imagery long into the night. I think I had four orgasms that night, over about three hours' time —a more intense and concentrated burst of sexual activity than I had ever had before, or than I have ever had since. One of the books that especially excited me that night was *Just like a Woman* [Thomas, 1989]. In this book an investigative reporter is feminized and transformed at the "Chrissy Institute." The feminization process at the fictional Institute involves hormones, electrolysis, and a diet and exercise regimen, as well as makeup, women's clothing, and training in feminine deportment [gender femaling co-opted in the service of erotic body femaling]. Some nine years later, and over three years after undergoing penectomy and vaginoplasty, I revisited *Just like a Woman*. I reread it, and found that the imagery still spoke to me erotically. I have continued to use *Just like a Woman* as a stimulus during masturbation, although these days it serves mostly as a jumping off point for my own fantasies. (personal communication to RE, 2000)

Very different are the migrating stories of those who place the primacy on gender migration, underplaying the erotic component to the point of extinction and only secondarily engaging in body migrating. The "high priestess" of gender migrating is Virginia Prince. In a series of publications extending over some 40 years (e.g., Prince, 1967, 1979, 1997) she has argued for a position that she came to term "transgenderist" (Prince, 1979:172). The "transgenderist" male wishes to express the feminine side of his personality. To do so, maximally, he lives full-time in the social role of a woman and dresses and comports himself appropriately. Cross-dressing is not an erotic experience. Any body femaling he engages in is done in the service of gender femaling. Unsightly body hair may be removed; the ingestion of hormones may lead to body feminization; but there is no sex-reassignment surgery—that would be an irrelevancy. Becoming the "other" gender on a permanent basis, and presenting as a woman socially (gender migrating) is the goal.

Oscillating Stories

As we saw in the quotation from McCloskey (1999), most migrating stories entail a one-way journey. What we are terming oscillating stories, in contrast, entail moving to and fro between male and female polarities, across and between the binary divide. The oscillators, in McCloskey's terms, are those who "try out the other gender for a day or a week or a month." We might say that the oscillator has a return ticket although the frequency of the journey and the length of stay will vary, depending on the transgenderer's social circumstances and his/her personal project. Some may spend a few hours every week on the other side of the divide; others may only manage the journey every few months but they might be able to stay for a week or more. There are oscillators who oscillate (largely in fantasy) minute to minute, even second to second. Oscillating stories may, like migrating stories, be seen as modernist tales. There are elements of silent suffering, discovery, and coming to terms with "being different," but these elements are less evident than in the migrating stories. In oscillating stories, we find less material on "being"—on identity and relationships—and more detail about "doing"—on excursions "over the gender border." As with the migrating stories, the binary divide is accepted. Medical help is rarely enlisted, although other people may play a part in facilitating the excursions.

The main sub-processes involved in oscillating stories are erasing, implying, concealing, and redefining. Except in fantasy, substituting is mainly, but not entirely, restricted to reversible substitutions and to those that can be concealed, such as males removing body hair. Oscillating stories will typically involve reference to the wearing of wigs, the use of padded bras and padded girdles, and the wearing of false moustaches and beards and something in the underpants to imply the possession of a penis. One common oscillating story is that of the male transvestite. Unlike migrating stories, it is rare to find oscillating stories in published autobiographies. Pepper (1982) and Rowe (1997) are exceptions. Often found in older medical case reports and occasionally in magazine feature articles, the most detailed stories are found in the newsletters of transvestite organizations such as the U.K. Beaumont Society. The following examples from research interviews illustrate some of the variations.

Pauline is 41, divorced, and living alone. She has a well paid job and lives in a very private, moderately expensive apartment. She never cross-dresses in public, so Pauline spends as much time as she can in her apartment, dressed as a woman, only leaving it for necessary work or shopping trips. She buys her large collection of women's clothes through mail order catalogs, and buys her wigs, padded girdles, and silicone breasts from a specialist mail order transvestite supplier. Her body is kept free of hair by depilating, shaving, or plucking. She was clean shaven at the time of interview, and was keenly experimenting with her new electrolysis machine.

Helena, on the other hand, participates in a number of social worlds within which she wishes to present different aspects of her oscillating. Helena is 35, single, and living alone. She advertises her services as a transvestite prostitute in contact magazines and for these engagements presents as a look-alike of 1940s movie stars. She maintains an hourglass figure with the use of a corset and a padded bra placed over her chest, carefully taped to create cleavage. Helena is also a member of a transvestite group that meets to cross-dress, principally in leather fetish gear. For this group, she adopts a punk image and redefines her slim male body as an anorexic girl's body. As a male, Helena works as a teacher in a boys' school, involving herself in cricket coaching in the summer months. The extent of her body erasing/substituting is limited primarily by her need to avoid possible detection in this work setting. In consequence, she never removes the hair from her hands and from the first two inches of her forearm, and throughout the cricket season leaves all her arm hair intact so that she can roll up her sleeves while bowling. In her cross-dressing, she makes a virtue out of necessity and extols the virtues of gloves—1940s and elegant-style for her work as a prostitute; 1980s and fetish-style for her punk look; and, in both cases, elbow length for the summer months.

In the above examples, substituting is limited by the need to return to the male side at some point. Pauline appears as a man in the relatively impersonal public world of work and is therefore able to substitute only in intimate ways that will not be outwardly visible. Helena substitutes differently according to the setting and time of the year. Both, in different settings, need or want to present as men but also as masculine, heterosexual non-transvestite men. Where this is not the case, substituting is likely to be more extensive and visible, as, for instance, where the oscillator is known to be a drag king (Volcano and Halberstam, 1999) or queen, prominent in gay and transvestite settings.

Fantasy rapid oscillating transgendering provides particularly illuminating illustrative material on the complex interrelations between sex, sexuality, and gender. John lives and works as a male. At home, he lives with his female partner, who sees herself as heterosexual. Before John joins his partner in bed at night, he applies a toner and a hormone bust enlargement cream to his nipples (preparatory substituting). He is not sure that the cream is leading to breast development, but it is making his nipples more sensitive. Occasionally he supplements this regimen with a course of hormone pills that he has acquired from a prostitute he visits. These do lead to breast tissue enlargement and sensitivity (embryonic substituting). He joins his wife in bed. As he touches her body he feels heterosexually aroused and with his penis erect enters his wife. For this period he identifies as a man with a man's body. In due time his wife climaxes. To maintain his erection, he now oscillates into female mode. He disavows his male body and "his" penis becomes his wife's. He now has his wife's vagina (fantasy substituting). When he climaxes, it is, he feels, a female orgasm. His pre-come is "her" lubrication. His ejaculatory fluid is his wife's. With intercourse over, his male self slowly reemerges.

In John's case his transgendering leads to what he fantasizes are female sexual responses. Indeed, his fantasy transgendering is a precondition of sustained sexual arousal and ejaculation. For others, like Peter, gender male femaling (i.e., actual cross-dressing) is a precondition of sexual arousal and ejaculation. Peter can only make love with his partner when he is cross-dressed, and although his partner does not encourage this male femaling, she goes along with it rather than lose him. Peter identifies as a transvestite with transsexual leanings. He has never had contact with medical professionals. Rather, his exploration of cross-dressing sub-cultures has led him to the view that his "oscillating" lifestyle is his best compromise. He has no interest in politicizing his position, and he keeps his transgendering practices and aspirations as private as possible: an oscillation between the boundaried dual worlds separated by the binary divide.

Negating Stories

The acknowledgment of "negating" as a conceptually distinct mode of transgendering is in its infancy. In the main, it is only within certain sections of the contemporary transgendered community that it has emerged as an identifiably different story.

Negating has two principal meanings: to nullify, make ineffective, invalidate, expunge, wipe out, cancel; and to deny, deny the existence of. Those negating negate in both senses. Those with male bodies seek to nullify their maleness and eliminate in themselves the existence of a binary divide. The objective is similar, but converse, with female negators. There is, of course, a literature on the eunuch: the castrated man. Stories of the eunuch may appropriately be classified as negating stories. Our research to date, however, has focused on contemporary Western negating, in particular, on contemporary male femaling "sissies" and, most recently, on life history work with the "ungendered," "gender-freeing" writer and activist, Christie Elan-Cane.

There are two principal variants of the negating story depending on the position of the negator in relation to the traditional version of the binary divide. In one variant, negating self-consciously buttresses the traditional version of the binary male/female divide for all males and females who are not negating/being negated. In the other variant, the person negating/being negated considers the binary divide a source of oppression and seeks a modification of the binary gender system that will enable those "without gender" to live "gender-free" lives.

Illustrative of the first variant are the writings of Debra Rose (1993a, 1993b, 1995a). Rose's writings are a blend of fact and fiction. They do, however, provide a particularly systematic account of one major "sissy maid" lifestyle and a lifestyle that has a considerable following. Many of our sissy informants tell us that her journal *Sissy Maid Quarterly* (SMQ, 1994–1996) "plugs into" their innermost thoughts, fantasies, and wishes with astonishing insight and accuracy.

Becoming a sissy maid entails taking on the duties of housemaid and personal maid to a mistress, either permanently or temporarily, and living for greater or lesser periods of time as a "neither male, nor female" sissy servant. This will entail systematic "sissification," either by the mistress that employs the maid, or by a "training" mistress, prior to service. In particular, the sissy maid's male sex, sexuality and gender are systematically erased.

In Rose's world, being male involves being masculine, active, virile and—by virtue of maleness—attracted to females; being female involves being feminine, being attractive to males, and finding males attractive because of their virility and masculinity. In short, it is a world where the congruity between sex, heterosexuality, and presentation of gender is taken for granted. There is no space for homosexuality, feminine men, or masculine women. In particular, heterosexuality and success in dating are necessary conditions of manhood and womanhood. And herein lies the problem for the sissy. Quite unable to compete for a female's affections in the conventional way, he must come to accept that he is neither male nor female but a "third gender" (Rose, 1995b:2). To secure this outcome, his already weak maleness must be systematically expunged.

While the details of the trainings vary, prominent in all trainings is systematic erasing. In the furtherance of erasing, there is much concealing (of attributes of masculinity) and much redefining (sissies must dress in effeminate and feminine attire *not* in order to cross the binary divide, but to further their emasculation). Such substituting and implying as there is, is co-opted in the service of erasing.

The sissy's body will be stripped of his body hair giving him a pre-pubescent look. Much attention will be paid to the practice of "gaffing." As Rose puts it: "Being gaffed is a very basic condition for a sissy male, with psychological as well as physiological benefits. By firmly pinning back his small, sissyish appendage, a gaff allows a sissy's appearance to more closely match his personality and inner (and ineffectual) sexuality" (Rose, 1994a:37). Once shaved and gaffed, the sissy is now ready to be "re-gendered" (ungendered). His male attire is replaced, initially by gender neutral clothes and later by clothes appropriate to his sissy-maid servant status. These will, most usually be women's clothes, but women's clothing redefined as male *sissy* maid clothing. Underwear may be emblazoned with sissy logos such as "never last." Rose (1994a:37) favors lycra tights or "bike" shorts because of the way these tight garments hug a gaffed sissy's front and function as a visual reminder of the male maid's sissiness.

Suitably depilated, gaffed, and attired, the sissy maid is now prepared for service: ready to take on the role of the housemaid. In addition, he will often act as personal maid to his mistress. Within this relationship, what might be termed the paradox of sissy sexuality is played out. Close to his mistress and privy to the intimacies of her boudoir and personal relationships, the residues of his unsatisfied and unsatisfiable heterosexual yearnings for her will be displaced from being genitally focused on her (in yearning and fantasy) to being suffused over his entire body in the sensual pleasure he obtains from the fit and texture of his effeminate clothing and the pleasures he obtains from the subservience of working for her.

Although he is permitted to masturbate, the elimination of his masculine sexuality will be enhanced by a particular form of masturbation deemed appropriate in the Debra Rose vision of sissy maid sexuality. Rose makes much of the pink rubber "sissy sheets" fitted to the sissy maid's single bed (e.g., Rose, 1994b). In time, many sissies will simply rub themselves "off" against the rubber sheets, focused on the pleasurable sensation, with no heterosexual (or homosexual) fantasy,[3] a semi-flaccid penis, and a weak dribble of an ejaculation (Rose, 1993b:1, 42). In due course, many sissies cease to masturbate entirely. The sissy maid becomes an increasingly asexual "neither male nor female" maid in service.

In marked contrast to Rose's vision of erasing is that of writer and activist Christie Elan-Cane. Christie is the first to pioneer publicly "per"[4] particular "ungendering." Per chosen terminology for per gender identity has shifted over the years from androgyne, through "third gender" (Elan-Cane, 1999a) and "gender free," to the now preferred "ungendered" or "non-gendered" (Elan-Cane, 1999b).

Christie, a biological female, felt "in the wrong body" from per earliest days. With the onset of puberty, per development of breasts and per menstruation was particularly distressing. As Christie (Elan-Cane, 1997:1) puts it: "I was never able to come to terms with 'womanhood.' I had the body of a woman and therefore I was considered by everyone to be a woman and I was repelled. I was disgusted by the physical changes to my body when I started to develop at puberty." Never identifying as transsexual, Christie eventually found a surgeon who would remove per breasts and, years later, per womb. Now at ease with per body, Christie began to identify as "neither male, nor female": "a third gender person." Although open to the possibility of further body erasing—having per ovaries removed, for instance—for Christie, "everything fell into place" once surgery was completed. Whereas Christie had flirted with a lesbian identity prior to surgery, now Christie found perself attracted to men, provided they saw per as being a third gender person. Per sexual partner is now a male who relates to per as non-gendered—neither male, nor female. Christie wears androgynous "gender-free" clothes—mostly black "neither male, nor female" shirts; black trouser-suits—a sort of contemporary Chairman Mao suit, and keeps per head meticulously shaven—all appropriate to per "ungendered" identification, following per body erasing. Having erased all undesired vestiges of femaleness and femininity, Christie has no need of further concealing, redefining, substituting, or implying.

Secure in per personal negating, Christie has increasingly turned per attention to publicizing per position, with the intention of enabling a space to be provided within a bi-polarized gender system for "ungendered" people like perself. Initial steps include campaigning for the use of a non-gender-specific form of address ("per" —derived from person; Pr. as a title to replace Mr. or Mrs.), and the inclusion of the gender-free in equal rights legislation (Elan-Cane 1997, 1998). In Christie's view "there is no reason why there should be two diametrically opposed genders nor is there any reason why gender should exist at all" (Elan-Cane, 1998:7) and, in these senses, per position might be seen as a subversive one. However, it is not per present intention to seek to undermine the binary divide, an approach that Christie regards as both unrealistic and impracticable. Rather, Christie seeks social legitimacy for perself and others like per (personal communication, 1998).

Transcending Stories

In recent years there have emerged a number of stories that entail rendering problematic the binary gender divide, in a way that puts the emphasis upon multiple and fluid genders (Bornstein, 1994; Feinberg, 1996; Halberstam, 1998). These stories are tales of the unexpected, par excellence. In this section we focus on these tales under the general title of "transcending stories." Whilst transgendering may be taking place by means of substituting, concealing, and implying, as in the other modes, the meaning of the transgendering sub-processes is fundamentally redefined.

Transcending stories are stories whose time has come. They are a part of what Plummer calls "the rise of the 'late modernist sexual story'" (1995:133) and they share their characteristics with stories in other fields. These stories do not replace modernist ones but rather coexist with them. Plummer identifies three main attributes of late modern stories: firstly, stories of authority give way to participant stories; secondly, stories of essence and truth give way to stories of difference; and thirdly, stories of the "categorically clear" give way to stories of deconstruction. We can discern these elements in transcending stories as medical authority is questioned, diversity is celebrated, and the certainty of sex and gender categories is called into question.

The style of the transcending story is also very different from that of the other three major modes. Transcending stories are not chronological personal narratives. The book-cover "blurb" for perhaps the best known transcending story, Kate Bornstein's *Gender Outlaw* (1994), describes it as "a manifesto, a memoir and a performance all rolled into one." In fact such memoirs are principally used as a vehicle to make questionable our "common-sense" assumptions about what it means to be a man or a woman. Similarly, in *Transgender Warriors* (1996), Leslie Feinberg uses his personal experiences to lead us through a transgender history and to outline his political philosophy. In *Read My Lips* (1997), Riki Wilchins mixes together personal experiences with current gender theories and transgender politics. The linearity of the modern story is replaced by "a little bit from here, a little bit from there. Sort of a cut-and-paste thing" (Bornstein, 1994:3).

In contrast to migrating stories—modernist tales of suffering and survival—these transcending stories tell bold tales of "transgender warriors" fighting a war against an enemy and for a people. The enemy becomes the cultural rules concerning gender, such as those discerned by Garfinkel (1967) in his study of Agnes. The people who are being fought for are the members of a broadly defined transgender community. The new transcending personal narratives convey a strong sense of being a part of a community, a movement organized around political action. In earlier migrating narratives the uniqueness of the story is often emphasized, as, for instance, in Cowell (1954) and Morris (1975). When previous "migrants" are mentioned in these stories, they are depicted as "remarkable," rare, and isolated individuals (Cowell, 1954:122–131).

Interwoven within late modernist tales is the emergence of a trans identity that is both permanent and "out" (Bolin, 1994:472–473). In a major variant of the migrating story, successfully claiming "transsexual" status is necessary in order to become entitled to hormonal and surgical body change, but it hardly functions as a central and permanent identity. The identity is female or male: transsexual denotes a temporary status. In a major variant of the oscillating story, "transvestite" may be a self-defined, more or less permanent identity, but it has largely been a secret one and its pathological connotations hardly a source of pride. In transcending stories, in contrast, transgender has emerged as an identity in itself, an alternative to an unambiguous male or female identity and one of which to be proud. As Bolin (1994:473) puts it: "it is akin to a new kind of ethnicity." So now, for some people, "the experience of crossed or transposed gender is a strong part of their gender identity; being out of the closet is part of that expression" (Nataf, 1996:16).

These stories then tell of battles against violence and discrimination and for various rights. The 1995 International Bill of Gender Rights (reprinted in Feinberg, 1996:171–175) claims that "all human beings have the right to define their own gender identity...to free expression of their self-defined gender identity" and to change "their bodies cosmetically, chemically, or surgically, so as to express a self-defined gender identity" (172–173).

This is contrary to the dominant view that the chemical and surgical alteration of the body is not a person's right but is only to be undertaken if authorized by an appropriate medical professional. The Harry Benjamin International Gender Dysphoria Association (1998), for instance, states in its Standards of Care for Gender Identity Disorders that, "hormones are not to be administered simply because patients demand them" (33) and that "surgical treatment for a person with a gender identity disorder [GID] is not merely another elective procedure. Typical elective procedures only involve a private mutually consenting contract between a suffering person and a technically competent surgeon. Surgeries for GID are to be undertaken only after a comprehensive evaluation by a qualified mental health professional" (37).

The idea of the "gender outlaw" shifts our attention to another way in which transcending stories go beyond our conventional understandings. This idea points to the position of transpeople as located somewhere outside the spaces customarily offered to men and women, as people who are beyond the laws of gender. So the assumption that there are only two (opposite) genders is opened up to scrutiny. Instead, the possibility of a "third" space outside the gender dichotomy is suggested.

A recent variant of the transcending story is being told by those people born with intersexed bodies. As Fausto-Sterling (1993) states, "Hermaphrodites have unruly bodies. They do not fall naturally into a binary classification; only a surgical shoehorn can put them there." And this is exactly what has happened: during the twentieth century, intersexed bodies have been surgically and hormonally fitted into one or the other gender category. Now, increasingly, people with intersexed bodies who were neither aware nor able to control such surgical and hormonal intervention are questioning those practices and demanding the right to determine whether, when, and how their bodies should be altered. Particularly notable, in this regard, is the work of Cheryl Chase, an intersexed woman and founder of the Intersex Society of North America (ISNA). Kessler (1998:79) writes:

> Simultaneous to ISNA's development, support groups for intersexuals were forming in Canada, Europe, Asia, Australia, Japan and New Zealand. In Germany, a group of intersexuals, using some of the same strategies as ISNA, established a peer support and advocacy group. The initial name of the group, Intersex Support Network Central Europe, was later changed to Genital Mutilation Survivor's Support Network and Workgroup on Violence in Pediatrics and Gynecology, reflecting the fury of its political evolution.

These intersex stories contain many of the elements of those transcending stories considered above: the emphasis on personal choice, the challenging of medical authority, the acceptance of bodies which are not unambiguously male or female.

In exploring transgender diversity through the personal narrative, these "tales of the unexpected" document the diversity of contemporary "options" for the expression of sex, sexuality, and gender. As Denny (1995:1) asserts:

> With the new way of looking at things, suddenly all sorts of options have opened up for transgendered people: living full-time without genital surgery, recreating in one gender role while working in another, identifying as neither gender, or both, blending characteristics of different genders in new and creative ways, identifying as genders and sexes heretofore

undreamed of—even designer genitals do not seem beyond reason.

References

Blanchard, R. (1989). The concept of autogynephilia and the typology of male gender dysphoria. *Journal of Nervous and Mental Disease*, Vol. 177: 616–623.

Blanchard, R. (1993). Partial versus complete autogynephilia and gender dysphoria. *Journal of Sex & Marital Therapy*, Vol. 19: 301–307.

Bolin, A. (1994). Transcending and transgendering: Male-to-female transsexuals, dichotomy and diversity. In G. Herdt (ed.), *Third sex, third gender: Beyond sexual dimorphism in culture and history*. New York: Zone Books.

Bornstein, K. (1994). *Gender outlaw: On men, women and the rest of us*. London: Routledge.

Concise Oxford Dictionary of current English (1995), 9th ed. Oxford: Oxford University Press.

Cowell, R. (1954). *Roberta Cowell's story*. London: Heinemann.

Denny, D. (1995). The paradigm shift is here! *Aegis News*, Vol. 4:1.

Ekins, R. (1983). The assignment of motives as a problem in the double hermeneutic: The case of transvestism and transsexuality. Paper presented at the Sociological Association of Ireland Annual Conference, Wexford, Ireland.

Ekins, R. (1984). Facets of femaling in some relations between sex, sexuality and gender. Paper presented at the Sociological Association of Ireland Annual Conference, Drogheda, Ireland.

Ekins, R. (1993). On Male femaling: A grounded theory approach to cross-dressing and sex-changing. *Sociological Review*, Vol. 41, No. 1: 1–29.

Ekins, R. (1997). *Male femaling: A grounded theory approach to cross-dressing and sex-changing*. London: Routledge.

Ekins, R., and D. King (eds.). (1996a). *Blending genders: Social aspects of cross-dressing and sex-changing*. London: Routledge.

Ekins, R., and D. King. (1996b). Is the future transgendered? In A. Purnell (ed.), *Proceedings of the 4th International Gender Dysphoria Conference*. London: Gender Trust, pp. 97–103.

Ekins, R., and D. King. (1999). Towards a sociology of transgendered bodies. *Sociological Review*, Vol. 47, No. 3:580–602.

Ekins, R., and D. King. (2000). Telling body transgendering stories. In K. Backett-Milburn and L. McKie (eds.). *Constructing gendered bodies*. London: Macmillan.

Elan-Cane, C. (1997). Prepared speech for cybergender discussion. Transgender Film and Video Festival, London.

Elan-Cane, C. (1998). A world without gender. Talk given at the Third International Congress on Sex and Gender, Exeter College, Oxford.

Elan-Cane, C. (1999a). Christie Elan-Cane. In T. O'Keefe (ed. K. Fox), *Sex, gender and sexuality: 21st century transformations*. London: Extraordinary People Press.

Elan-Cane, C. (1999b). A life without gender in a gendered society. Paper presented at the 16th Harry Benjamin International Gender Dysphoria Association Symposium, London.

Fallowell, D., and A. Ashley. (1982). *April Ashley's Odyssey*. London: Jonathan

Cape.
Fausto-Sterling, A. (1993). The five sexes: Why male and female are not enough. *Sciences*, Vol. 33, No. 2: 20–25.
Feinberg, L. (1996). *Transgender warriors: Making history from Joan of Arc to Dennis Rodman.* Boston: Beacon Press.
Garfinkel, H. (1967). *Studies in ethnomethodology.* Englewood Cliffs, NJ: Prentice Hall.
Halberstam, J. (1998). *Female masculinity.* Durham, NC: Duke University Press.
Kessler, Suzanne. (1998). *Lessons from the intersexed.* New Brunswick, NJ: Rutgers University Press.
Kessler, S. J., and W. McKenna. (1978). *Gender: An ethnomethodological approach.* New York: Wiley.
King, D. (1981). Gender confusions: Psychological and psychiatric conceptions of transvestism and transsexuality. In K. Plummer (ed.), *The making of the modern homosexual.* London: Hutchinson.
King, D. (1993). *The transvestite and the transsexual: Public categories and private identities.* Aldershot, England: Avebury.
Lawrence, A. (1999). Men trapped in men's bodies: Autogynephilic eroticism as a motive for seeking sex reassignment. Paper presented at the 16th Harry Benjamin International Gender Dysphoria Association Symposium, London.
McCloskey, D. (1999). *Crossing: A memoir.* Chicago: University of Chicago Press.
Mills, C. Wright. (1940). Situated actions and vocabularies of motive. In I. L. Horowitz. (ed.), *Power, politics and people: The collected essays of C. Wright Mills.* New York: Oxford University Press.
Morris, J. (1975). *Conundrum.* London: Coronet Books.
Nataf, Z. I. (1996). *Lesbians talk transgender.* London: Scarlet Press.
Pepper, J. (1982). *A man's tale.* London: Quartet Books.
Plummer, K. (1995). *Telling sexual stories: Power, change and social worlds.* London: Routledge.
Plummer, K. (1996). Intimate citizenship and the culture of sexual story telling. In J. Weeks and J. Holland (eds.), *Sexual cultures: Communities, values and intimacy.* London: Macmillan.
Prince, V. (1967). *The transvestite and his wife.* Los Angeles: Chevalier Publications.
Prince, V. (1979). Charles to Virginia: Sex research as a personal experience. In V. Bullough (ed.), *The frontiers of sex research: The latest findings.* Amherst, NY: Prometheus Books.
Prince, V. (1997). My accidental career. In B. Bullough, V. Bullough, M. Fithian, W. Hartman, and R. Klein (eds.). *How I got into sex.* Amherst, NY: Prometheus Books.
Ramet, S. (1996). *Gender reversals and gender cultures.* London: Routledge.
Rose, D. R. (1993a). *Maid in form "A", "B", and "C".* Capistrano Beach, CA: Sandy Thomas.
Rose, D. R. (1993b). *The Sissy Maid Academy.* Vols. 1 & 2. Capistrano Beach, CA: Sandy Thomas.
Rose, D. R. (1994a). Top drawer. *Sissy Maid Quarterly*, Vol. 1, 36–37.
Rose, D. R. (1994b). The Maid's room—we test rubber sheets. *Sissy Maid Quarterly*,

Vol. 2:27-30.

Rose, D. R. (1995a). *Where the sissies come from.* Capistrano Beach, CA: Sandy Thomas.

Rose, D. R. (1995b). Acceptance—The key to happiness in the maid's room. *Sissy Maid Quarterly*, Vol. 3: 2-5.

Rowe, R. (1997). *Bert & Lori: The autobiography of a crossdresser*, Amherst, NY: Prometheus Books.

SMQ (1994-1996), Sissy Maid Quarterly, Numbers 1-5. California: A Sandy Thomas Publication, produced in conjunction with Rose Productions.

Spry, J. (1997). *Orlando's sleep—An autobiography of gender.* Norwich, VT: New Victoria Publishers.

Thomas, S. (1989). *Just like a woman.* Capistrano Beach, CA: Sandy Thomas.

Thompson, R., with K. Sewell. (1995). *What took you so long: A girl's journey to manhood.* London: Penguin.

Volcano, D., and J. Halberstam. (1999). *The drag king book.* London: Serpent's Tale.

von Mahlsdorf, C. (1995). *I am my own woman: The outlaw life of Charlotte von Mahlsdorf.* Pittsburgh: Cleis Press.

Webb, T. (1996). Autobiographical fragments from a transsexual activist. In R. Ekins and D. King (eds.), *Blending genders: Social aspects of cross-dressing and sex-changing.* London: Routledge.

Wilchins, R. (1997). *Read my lips: Sexual subversion and the end of gender.* Ithaca, NY: Firebrand Books.

Notes

We wish to extend our warmest appreciation to all those informants who have shared their stories with us, whether personally identified in this chapter or included under another name. Sections of a number of the narratives included in the chapter have appeared in Ekins and King (1999, 2000). These passages are reproduced by courtesy of the *Sociological Review* and Macmillan Press. Special thanks go to Wendy Saunderson for her help in bringing the chapter to final fruition.

1. See Webb (1996). Rachael—then writing as Terri—now wishes to disassociate herself from many of the views she expressed in this 1996 chapter.

2. Whereas, in the main, it is uncontroversial to consider "transvestism" in terms of its erotic (autogynephilic) components, to do likewise with "transsexualism" is controversial. Blanchard's extended use of the term and its hostile reception within sections of the transgender community warrant detailed study.

3. Sissy Diana, for instance, in the list he has made of "specific attitudes I must change to become more like a proper sissy," writes: "If I play with myself at night I must only think of what a total sissy I am" (personal communication, 2000).

4. For Christie Elan-Cane: "To gain validity and social legitimacy it is imperative that the Third Gender has a proper title that is non-gender specific and a correct form of address." Christie's own preference rejects "his" or "her" for "per," derived from "person" (Elan-Cane, 1998:3-4).

Studying Transsexual Identity

Katherine Johnson

Theoretical accounts in the field of transsexualism can be broadly classified into two major categories. The first is clinical in its approach, encompassing medical, psychiatric, and psychological research that perceives transsexualism as a syndrome, subject to treatment and observation. The second approach has a more sociocultural outlook, in that it is principally concerned with the relationship transsexualism has to the culture at large (Bolin, 1988). However, whether focusing on the etiology (Stoller, 1973) or analyzing transsexualism in a quest to understand how gender is socially constructed (Kessler and McKenna, 1985), transsexuals have been theorized and studied in an effort to understand the broader categories of sex and gender.

Traditionally gender has been used to distinguish psychological, social and cultural aspects of maleness and femaleness (Kessler and McKenna, 1985). Within this context, it is now widely agreed that there are two sexes, defined in terms of the anatomical body—male and female—and two genders, defined in terms of personality traits and behaviors—masculine or feminine—that are often seen as socially constructed. Yet, in most attempts to distinguish between "sex" and "gender," there is an assumption that "sex" is somehow prior to gender—the biological bedrock upon which gender differences are constructed (Kitzinger, 1994). Alternative perspectives are provided, however, particularly by those working within a post-structuralist or Foucauldian framework. These studies challenge the "truth" of the Western belief that there are two, and only two, biological sexes, suggesting that the "reality effect" of this belief has been produced by powerful medical and scientific discourses (see Butler, 1990; Laqueur, 1990).

The term "transsexualism" was first coined by Harry Benjamin (1953). It is unusual because it names the method of treatment and rehabilitation rather than the syndrome—a syndrome for which there is no known cause. Without a known cause, transsexualism is open to self-definition through personal suffering. Physicians, however, have constructed their own criteria of symptoms that an individual must exhibit before they will accept any candidate's self-defined condition. From an extensive literature review, Roberto (1983:17) concluded that the symptoms required for a clinical definition of transsexualism included

> the belief that one is a member of the opposite sex...dressing and behaving in the opposite gender role...perceiving oneself as heterosexual although sexual partners are anatomically identical...repugnance for one's own genitals and the wish to transform them...and a persistent desire for conversion surgery.

In fact, the demand for surgery is often regarded as inherent in the clinical definition of transsexualism (Hoenig, 1982). Thus, the clinical literature may attempt to provide a classification system or an etiologic basis for transsexualism, but at the same time it also creates "meanings" or "stories" (Plummer, 1995) that qualify as constituting the transsexual experience. However, this is not the only genre where such self-construction takes place. A transsexual identity is also constructed by the discourses that evolve from others who have gone through the same reassignment process. As transsexual individuals' own accounts become more available, for example, through autobiography, "subjectivity" is becoming an increasingly recognized site for the meeting of various discourses. Here, the term subjectivity is used in the same sense as outlined by the authors of *Changing the Subject: Psychology, Social Regulation and Subjectivity*:

> We use "subjectivity" to refer to individuality and self-awareness—the condition of being a subject—but understand in this usage that subjects are dynamic and multiple, always positioned in relation to discourses and practices and produced by these—the condition of being a subject. (Henriques et al., 1984:3)

Furthermore, through newspaper articles and television documentaries, media coverage provides a third site where discourses concerning the nature and status of transsexualism can circulate between academic literature, transsexual individual's subjective accounts, and lay interpretations.

The purpose of this chapter is to formulate a methodological approach that can be utilized to trace the discursive constructions that shape and mold transsexual identity. A theoretical framework for the study of identity will be outlined, before highlighting the poststructuralist assumptions that underpin it. Two forms of discourse analysis employed within the discipline of psychology will be discussed, while concluding comments will consider some criticisms and methodological developments for future incorporation.

The term "identity" is increasingly used in conjunction with variously defined groups, whether they be youths, gays or transsexuals. Given the prolific use of the term "identity," it is important to specify exactly what one means by identity. Therefore, an approach originally outlined by Kathryn Woodward (1997) will be drawn on to illustrate the stance taken in this chapter. Firstly, this framework incorporates a shift away from essentialist notions of belonging, instead, focusing on non-essentialist definitions of identity, which highlight difference as well as common or shared characteristics. For example, if we look back at the definition of transsexualism provided in the quotation from Roberto (1983), we see a criterion for transition that is effective in molding a very fixed and rigid account of transsexual experience. This stands in stark comparison to an excerpt taken from a transgender community leaflet:

> Some (but not all) transgender people may use hormones or surgery, but these have serious effects on general health, mental and emotional well-being, and sexual pleasure and function. Many express their gender without medical aid, and most do not have genital surgery. Some identify as male, female, both (bi-gender), or neither; as male to female, female to male, transsexual, man or woman of transgender background, or person with transgender qualities; as gay, straight, lesbian, bisexual, heterosexual or other. However we may identify, and whatever medical options we choose, we all deserve to be treated with equal dignity and respect. (norrie mAy-welby, 1997)

The first account emerges from theorizing the transsexual condition within the traditional "medical model" (Kando, 1973), which involves the collection of biographical and in-depth psychological data followed by a period of analysis, classification, diagnosis, and etiological theorizing (Ekins, 1997). However, it may be necessary to challenge these types of accounts, and allow for diversity in experience, if we are to establish a greater understanding of why certain individuals position themselves within the discourses of transsexualism.

The second tenet of Woodward's theoretical framework is that an identity is always relational. An identity is formed, not necessarily in opposition to somebody else, but in relation to, or differing from, somebody else. This identity can be established through *symbolic marking*. For example, badges or uniforms are two methods of marking a specific identity. In the case of transsexual identity, wearing "Transsexual Menace" T-shirts has proved an effective means of symbolically marking a transsexual political identity. It is not crucial, however, that the marking of an identity be physically identifiable, because identity is also maintained through social and material conditions. For example, legal discourses mark transsexuals as "outsiders," as "different" or even taboo. In the U.K., transsexuals cannot entirely take up the position of male or female, and have no legal right to change their birth certificates, marry, and become the adoptive parents of their partner's children. Furthermore, since the case of "P. vs S. and Cornwall County Council" in which the European Court of Justice ruled that P's dismissal, because of her intention to seek gender reassignment, was unfair on the grounds of sex, the government has sort to introduce amendments to the Sex Discrimination Act of 1975 regarding cases of transsexualism. The proposed amendments focus on the interim period of six months to a year when the individual is in transition. For example:

> 14 c) During the period and six months afterwards, it will be lawful to exclude the individual from jobs involving contact with members of the public or customers who are changing e.g. staff in health clubs, clothes shop assistants, home helps, swimming attendants etc. (DfEE: Consultation Paper; February, 1998)

Whatever one's opinion about the validity of such proposals, these changes to employment legislation could have very real effects for the transsexual, including social exclusion and material disadvantages.

Finally, Woodward's framework professes that identities are not unified. There may be contradictions that have to be negotiated, both within the individual and between the individual and other group members. For example, contradictions may arise between the political identity, such as that required by Press for Change or Transsexual Menace, and the individual identities of those living in a shared culture. This point can be illustrated through a comparison of some of the visual images used to publicize Press for Change's political campaign, and images and artwork coming out of the transgender community. Press for Change, for good reason, presents the conservative, conformist face of transsexuality, while images such as those produced by Loren Cameron (1996) can be read as transgressive—a radical challenge to our gender assumptions.

This theoretical framework is informed by the principles of poststructuralist theory, which emphasizes that when conceptualizing identity it is necessary to attend to different dimensions, as well as highlighting difference and incorporating the notion that subjectivity is constituted through language.

Poststructuralism makes certain assumptions about language, subjectivity, knowledge, and truth (Weedon, 1987). Its founding principle is that our social reality is constituted through language, rather than reflecting a material world that is already given. There is no universally shared meaning for any given concept, as meaning can vary from language to language, from culture to culture, and across time. Therefore, a poststructuralist account assumes that meaning is composed within language and is not guaranteed by the subject that speaks it. Thus, language is paramount to the study of transsexual identity, as it is the place where a sense of self and subjectivity is constructed. Who we are and how we understand ourselves do not originate in "pre-packaged" forms inside us. Our identities are all effects of language, as the structure of language determines the way that experience and consciousness are structured (Woodward, 1997). Accordingly, given that language is a social phenomenon, identity construction cannot be accomplished outside of the social setting. Hence, the construction of identity is dependent on exchanges between people, whether this be through face-to-face dialogue or via the media of reading, radio, and television. An example taken from a recent interview with a male-to-female [MtF] transsexual clearly illustrates the central role of language in her process of transition:

KJ: How long did the reassignment take you?
Stella: ...umm...well I, *I'd realised from speaking to Dina exactly who I was*. That I wasn't a drag queen, I wasn't a transvestite, umm, that I was like she was. *I was actually transsexual.*

Language is clearly the crucible of change for her identity. It was only after talking with a friend, also a transsexual woman, that Stella realized who she was. She came across her identity as transsexual through language, in conversation with a friend. This point brings us to one of the most liberating aspects of this approach. If language is the place where identities are built and maintained, then poststructuralist theory sees language as the major site where oppressive identities can be challenged or changed. What it means to be a transsexual, woman, or man can be transformed and reconstructed. Language offers us various discursive positions through which we can consciously live our lives (Weedon, 1987). If, however, these language structures and discursive positions provide us with our notions of selfhood and personal identity, we need to formulate a method that enables us to identify them and examine them more closely.

Discourse Analysis: The Method

Psychological research traditionally exerts great effort to adhere to methods commonly found in the physical sciences. This raises questions and debates that are particularly pertinent to the study of mental states and human experience. Can scientists control all the possible variables that may be involved when considering human nature or behavior? Can the study of the self be reduced to a "true" entity? Here, "true" is placed in scare quotes as some researchers, usually those influenced by poststructuralist or postmodernist perspectives, question the notion that there really is a truth out there waiting to be discovered. These theorists draw, to varying degrees, on the assumptions of social constructionism which argues that knowledge is socially and culturally specific. Thus, in what has been documented as "the turn to language" (Parker, 1989), there has become an increasing trend for social scientists to include methods of discursive analysis within their research.

Michel Foucault (1969) defined discourses as "the practices that systematically form the objects of which we speak" (49), while others use it in its most open sense, to cover all forms of spoken interaction, formal and informal, and written texts of all kinds (Gilbert and Mulkay, 1984; Potter and Wetherell, 1987). Discourse analysis is almost synonymous with critical and in some cases feminist research. It provides us with a useful analytic tool that can be employed to identify and tease apart the discourses that are at work in a particular text. The findings can then be used, amongst other things, to comment on social processes which participate in the maintenance of structures of oppression. Undoubtedly, the most influential proponent of this approach emerging from the poststructuralist tradition was Foucault. In the light of his genealogies of power, notions of power and oppression have undergone major transformations. As Elizabeth Grosz (1995) points out, he has rendered the notion of oppression considerably more sophisticated by alerting us to the idea that the attribution of social value is not simply a matter of being depicted as passive and compliant, stripped of all forms of resistance. Rather, a position of subordination "exerts its own kind of forces...its own practices, and knowledges, which, depending on their socio-cultural placement and the contingencies of the power game that we have no choice but to continue playing, may be propelled into positions of power and domination" (Grosz, 1995:210). Accordingly, the work of Foucault has provided a sense of hope for those classified as oppressed, for there is always the possibility of a certain agency that will enable them both to challenge and to transform their position. In fact, Foucault's work has been drawn on heavily by many feminists to contest gender inequity, thus informing their political practices and struggles. Consequently, attention to discourse facilitates multiple goals, providing a historical account of knowledge, a critique of psychological/legal/medical practices by challenging their truth claims, and informed political practice and struggles (Burman and Parker, 1993).

Discourse analysis is used in many variants from cognitive linguistics to deconstruction, affording a means to analyze language and texts. In psychology, Discourse Analysis is now a well-established method (Bannister et al., 1994), but there are two distinct approaches, which Vivian Burr (1995) usefully distinguishes between as "discourse analysis" (e.g., Potter and Wetherell, 1987), and the "analysis of discourse" (e.g., Hollway, 1989; Parker, 1992).

Jonathon Potter and Margaret Wetherell's (1987) *Discourse and Social Psychology: Beyond Attitudes and Behaviour* is perhaps one of the most influential research methods publications in recent times, ushering in a new era for psychological research. Rather than using the term "discourse," they refer to "interpretative repertoires" which highlight the way an individual frames an issue. In a similar vein, they are looking to analyze the "performative qualities of discourses" in order to theorize what people are doing with their talk and writing, deciphering what they are trying to achieve. This notion of language function is one of the central suppositions of Discourse Analysis. However, the analysis of function cannot be seen in the simplistic terms of categorizing pieces of speech because it is very much dependent upon the analyst's "reading" of the context, in which the speech occurs. Furthermore, a person's account will often vary in accordance with its function or the purpose of the talk. This notion of variation is the second major component of discourse analysis. For discourse analysts' variation is as important as, if not more important than, consistency. This stands in stark contrast to the almost near-fetishism for limitation of variance within traditional quantitative research, where it is not uncommon for researchers relying on statistical methods to discard outliers, or those that vary too much, in an effort to return a significant result. But, as Potter and Wetherell point out, if talk is orientated to many different functions, any examination of language over time will reveal considerable variation. This variation occurs because, on every occasion, people are using their language to construct versions of the world. This leads us to the third crucial component of discourse analysis, the assumption that talk and writing is constructed out of preexisting resources. Inevitably, this will involve the exclusion of some resources. Reflecting on what has been left out is of as much value to the researcher as the analysis of what is included, because absences will aid in the construction of a particular version of "reality."

The principles that have been outlined here are common to most forms of discourse analysis. Where the approaches differ is in their consideration of subjectivity. For Potter and Wetherell, nothing exists outside of the text. The sole purpose of their analysis is to determine the action orientation of any given rhetoric. As they state:

> the researcher should bracket off the whole issue of the quality of accounts as accurate or inaccurate descriptions of mental states...Our focus is exclusively on discourse itself: how it is constructed, its functions and the consequences which arise from different discursive organization. (1987:178)

But in doing this, they completely bypass the notion of subjectivity. In fact, frequent criticism has been directed at this approach, occasionally charging Potter and Wetherell with returning to some form of behaviorism, where the mind is perceived only as a black box. Despite wishing to avoid these charges, they fail to move beyond acknowledging the limitations of "blank subjectivity" (Parker, 1997). Instinctively, this approach does not seem sufficient for those endeavoring to undertake psychological research, in which mental states, consciousness, or notions of how individuals experience their self, identity, and world are of utmost importance.

The second approach, the analysis of discourses (e.g., Parker 1992; Hollway, 1989), has developed out of attempts to solve the problem of theorizing subjectivity. Within this approach prevailing discourses of, for example, gender and sexuality are examined and their identity and power implications are bought to the fore. Often this involves identifying the positions offered by different discourses, spelling out the identity and political implications of these. It incorporates a Foucauldian approach, analyzing how specific discourses become grounded in social and material reality. Foucault, through his analysis of institutions such as the asylum, showed how discourses that originated in the institution are taken up by the subject, and then loop back to both legitimate and perpetuate the institution (Foucault, 1973). Drawing your attention back to the clinical classification of transsexualism (144), precisely this process took place in the early diagnostic and classificatory systems. It has been well documented that by the 1970s clinicians had become concerned about the lack of variation in prospective reassignment candidates' personal accounts (Stoller, 1973; Billings and Urban, 1982/1990). Having set up a rigid classification system, which needed to be adhered to, those requiring reassignment simply accessed the necessary documentation and presented themselves as "textbook" cases. Thus, the clinicians positioned transsexuals within specific discourses, which were then reinforced by those who, necessarily, positioned themselves in the same discourses when presenting themselves for reassignment. Therefore, this approach allows us to show how objects are constructed in discourses, but it also shows how subjects are constructed.

Wendy Hollway (1989) draws on the notions of "positioning" and "multiple subjectivity" to qualify the experience of being a subject. "Positioning" refers to how subjects are constructed through identifying their subjective experience within specific discourses, positioning themselves within them. Furthermore, she found that subjects position themselves within varying and often contradictory discourses, invoking the term "multiple subjectivity," and lending support to the poststructuralist attack on the Western philosophical notion of a rational and unified self. The image of the self as multiple challenges more traditional models, such as those proposed by trait theory, role theory, and humanistic accounts, which perceive the self as an entity, with one true nature that is waiting to be discovered. These theories have been paramount to psychology's conception of the individual person, making it possible to contrast the individual with society, as natural pairs in a balanced dichotomy (Potter and Wetherell, 1987). Moreover, under the encompassing heading of social constructionism, a whole new branch of critical psychology has emerged, dedicating considerable effort to reconceptualizing the subject (e.g., Gergen, 1985; Henriques et al., 1984). This incorporation of different linguistic practices into conceptualizing the self is a radical, political, and potentially emancipatory activity (Potter and Wetherell, 1987), because if subjectivity is constituted through language, then it allows for the possibility of change. As Hollway (1989:97) states:

> it is possible to transform the meaning of experience by bringing a different set assumptions to bear on it....in consciousness-raising groups, women learn to position themselves in a feminist discourse.

The problem with this approach is that the ability to change has to be seen as only a *potential* ability. We must remember that some positions are not easily transformed, and some people will hang on ferociously to particular discursive positions, even when they are consciously aware that they produce a negative, oppressed, or even destructive sense of self. As Lynne Segal perceptively writes in response to Judith Butler's (1990) notion of gender performativity:

> Mostly we can only enact those behaviours which have long since become familiar and meaningful to us in expressing ourselves...Challenge to our gendered "identities" may be more than we can handle. (Segal, 1994: 208).

Despite the potentially liberating qualities of this approach for theorizing subjectivity and identity, we are left in something of a quandary. If we view the subject only as a set of multiple and contradictory positionings or subjectivities, how can we account for the continuity of the subject and the subjective experience of identity? This is the shortcoming of a discursive approach and the principal area of contention—to only see the subject as bound up in [and the effect of] multiple and varying discourses does not provide all the necessary components for a comprehensive theory of identity.

Beyond Discourse: Criticisms and Developments

One of the most potent fears when employing social constructionist argument, is that critics will misuse them to the further detriment of an already oppressed group. Carole Vance (1984) clearly illustrated this in the case of homosexuality, describing how interpretations of social constructionist theory by right-wing proponents of "family values" interpretation of the theory led them to conclude that individuals had the ability to change at will or voluntarily. In a similar vein, this type of critique could be leveled against transsexuals. If gender identity is socially constructed, then they can change their identity, removing the need for hormone prescriptions, reassignment surgery, and legal recognition of their "new" gender. This is of course a gross misinterpretation of the theory, and as has already been pointed out, individuals often do not have the power to change their sexual and gender identities. Therefore, it is important to be aware that the discursive constitution of subjectivity is more than the individual consciously identifying with particular subject positions within a discourse. Rather, it has been argued that we need to move beyond discourse by attending to the *psychic* level as well as symbolic marking and social and material conditions, if we are to conceptualize a persuasive theory of identity (Segal, 1990, 1994; Woodward, 1997). In fact, Wendy Hollway (1989) drew on the psychoanalytic concepts of "splitting" and "repression" in order to explain why the participants in her research became attached to particular discursive positions. Moreover, we need to register that we are also placed in specific subject positions by others. This is especially important in the case of transsexuals. Inevitably, we will be allocated to the gendered position of male or female on the bases of others readings of our appearance and physicality.

Undoubtedly, the body plays a crucial role in the construction of an individual's self-identity, as the body acts as a symbolic marker of gender in all our social interactions. If a body is read or recognized as male, then subjects will be positioned within a discourse of masculinity, whether or not they would position themselves in that discourse. So, for MtF transsexuals, being read as male and positioned in the discourses of masculinity can be very distressing when they position themselves within a discourse of femininity. During recent years, we have witnessed a huge growth in the "technologization of the body" (Henriques et al., 1984), exemplified by practices ranging from weight training to gender reassignment, in which various techniques are used to shape and hone the body. The notion of an increasingly malleable body has introduced a new and exciting referent into the identity equation. To speak of subjectivity is simply no longer sufficient as it becomes ever more apparent that subjectivity is always already embodied. Despite the influence of technologies of the body, with their endless possibilities for the construction of new identities, there remain very many bodily and material limitations that science cannot transcend (Henriques et al., 1984). Phalloplasty, for example, is still an unviable option for many females to males [FtMs] given the relatively poor surgical outcomes. I believe that it is essential to develop a theory of embodied subjectivity, partly for studying the experience of living a certain subjectivity, but also because the body constrains our subjectivity. It is the body, through its fleshy physicality and the social processes that shape it, that restricts our ability to position ourselves within particular discourses or identify with specific discursive constructs. While the definitions, the day-to-day interpretations of manhood and womanhood, are so bound up with their biology, we cannot see our subjective experience of gender as anything but regulated and constrained by our physicality. It is therefore important to expand on a discursive approach to identity, which either tends to see the body as a vessel for the self, or completely overlooks it. We may need to develop our methodological tools if we are to work toward a theory of embodied subjectivity, where we attend to the subjective knowledge of transsexuals, but where we also recognize that this knowledge is embodied. As Henry Rubin wrote in a recent transgender anthology, arguing for the incorporation of a phenomenological approach in trans studies:

> it seems particularly prudent to use a method that not only legitimates subjectively informed knowledge but also recognizes the significance of bodies for the lived experience of the I. (Rubin, 1998:268).

Conclusion

Identity cannot be seen as anything but socially constructed, when the meanings that constitute identity are built, challenged, and transformed through language. Discourse analysis provides us with an effective methodological tool for teasing out the various discourses and discursive constructs that constitute an individual's experience. However, it is not a sufficient explanation to view the transsexual as simply an opportunistic manifestation of discourses, if we are to gain an insightful understanding of *why* certain individuals position themselves within the discourses of transsexualism. Instead, it is argued that this approach needs to be expanded by attending to the psychic level, as well as emphasizing that our subjectivity, our sense of self, and our gender identities are all housed within an anatomically and biologically defined body, a body that has very real effects on the regulation and constraint of our experience as gendered subjects.

References

Bannister, P., E. Burman, E. Parker, M. Taylor, C. Tindall. (1994). *Qualitative methods in psychology: A research guide.* Buckingham, England, and Philadelphia: Open University Press.

Benjamin, H. (1953). Transvestism and transsexualism. *International Journal of Sexology*, Vol. 7, 12–14.

Billings, D. B. and T. Urban. (1982/1990). The socio-medical construction of transsexualism: An interpretation and critique. In Richard Ekins and Dave King (eds.), *Blending Genders: Social Aspects of Cross Dressing and Sex-Changing.* London and New York: Routledge.

Bolin, A. (1988). *In search of Eve: Transsexual rites of passage.* South Hadley, MA: Bergin and Harvey.

Burman, E., and I. Parker. (1993). Discourse analysis: The turn to the text. In E. Burman and I. Parker (eds.), *Discourse analytic research. Repertoires and readings of texts in action.* London: Routledge.

Burr, V. (1995). *An introduction to social constructionism.* London: Routledge.

Butler, J. (1990). *Gender trouble: Feminism and the subversion of identity.* New York: Routledge.

Cameron, L. (1996). *Body alchemy: Transsexual portraits.* Pittsburgh: Cleis Press.

DfEE (Department for Education and Employment) Consultation Paper. (1998). *Legislation regarding discrimination on grounds of transsexualism in employment.* In *Press for Change Newsletter* (February, 1998).

Ekins, R. (1997). *Male femaling: A grounded theory approach to cross-dressing and sex-changing.* London: Routledge.

Foucault, M. (1969). *The archaeology of knowledge.* London: Tavistock.

Foucault, M. (1973). *The birth of the clinic.* Trans. A. M. Sheridan. London: Tavistock.

Gergen, K. J. (1985). Social constructionist inquiry: Context and implications. In K. J. Gergen and K. E. Davis (eds.), *The social construction of the person.* New York: Springer-Verlag.

Gilbert, G. N., and Mulkay, M. (1984). *Opening Pandora's box: A sociological*

analysis of scientists' discourse. Cambridge: Cambridge University Press.
Grosz, E. (1995). *Space, time and perversion.* London and New York: Routledge.
Henriques, J., W. Hollway, C. Urvin, C. Venn, and V. Walkerdine (eds.). (1984). *Changing the subject: Psychology, social regulation and subjectivity.* London: Methuen.
Hoenig, J. (1982). Transsexualism. In *Recent advances in clinical psychiatry,* Vol. 4. Ed. K. Granville-Grossman. Edinburgh: Churchill Livingstone.
Hollway, W. (1989). *Subjectivity and method in psychology: Gender meaning and science.* London: Sage.
Kando. (1973). *Sex change: The achievement of gender identity among feminized transsexuals.* Springfield, IL: Charles C. Thomas.
Kessler, S. and W. McKenna. (1985). *Gender: An ethnomethodological approach (2nd Ed.).* Chicago and London: University of Chicago Press.
Kitzinger, C. (1994). Editor's introduction: sex differences—Feminist perspectives. (Special Feature). *Feminism and Psychology,* 4, 501–506.
Laqueur, T. (1990). *Making sex: Body and gender from the Greeks to Freud.* Cambridge, MA: Harvard University Press.
mAy-welby, N. (1997). *A brief cartoon guide to gender and transgender.* Transnational Spansexual Foundation: *http://www.cat.org.au/ultra/sex.html*
Parker, I. (1989). *The crisis in modern social psychology, and how to end it.* London: Routledge.
Parker, I. (1992). *Discourse dynamics: Critical analysis for social and individual psychology.* London: Routledge.
Parker, I. (1997). Discourse analysis and psychoanalysis. *British Journal of Social Psychology,* 36, 479–495.
Plummer, K. (1995). *Telling sexual stories: Power, change and social worlds.* London; New York: Routledge.
Potter, J., and M. Wetherell. (1987). *Discourse and social psychology: Beyond attitudes and behaviour.* London: Sage.
Roberto, L. G. (1983). Issues in diagnosis and treatment of transsexualism. *Archives of Sexual Behaviour,* 12, 465–473.
Rubin, H. (1998). Phenomenology as method in trans studies. *GLQ: A Journal of Lesbian and Gay Studies,* 4, No. 2, 263–282.
Segal, L. (1990). *Slow motion: Changing men, changing masculinity.* London: Virago.
Segal, L. (1994). *Straight sex: The politics of pleasure.* London: Virago.
Stoller, R. J. (1973). Male transsexualism: Uneasiness. *American Journal of Psychiatry,* Vol. 22, 47–64.
Vance, C. (1984). Social Construction Theory: Problems in the history of sexuality. In H. Crawley and S. Himmelwait (eds.), *Knowing women: Feminism and knowledge.* Cambridge: Open University Press.
Weedon, C. (1987). *Feminist practice and poststructuralist theory.* Oxford: Basil Blackwell.
Woodward, K. (1997). Concepts of identity and difference. In K. Woodward (ed.), *Identity and difference.* London: Sage and Open University Press.

Gender Love and Gender Freedom

Surya Monro

The development of theoretical foundations for the conceptualization of "trans" is crucial for education and social policy concerning gender. This chapter outlines a contribution to such a framework by exploring the tensions between postmodernist and structural analysis of trans. A postmodernist model of trans usefully describes the fluidity, multiplicity, and paradox that can be found among some transsexual and transgender people. Gender politics based on discrete male-female categories logically becomes problematized by such identities as does a social structure based upon rigid gender binaries. However, the postmodernist model is inadequate in describing trans, as it fails to account for the trans sense of self or the impact of social structures upon the fluidity decried earlier. It is concluded that an analysis combining postmodern and structuralist accounts is necessary for understanding trans.

In conducting this research I took a participative, feminist approach. I am a woman born as a woman who challenges gender stereotypes in various ways. I triangulated my research methodologies, which consisted primarily of in-depth interviewing, participant observation, and e-mail discussion. In-depth interviewees numbered 22 and these included pre- and post-operative female-to-male and male-to-female transsexuals, transvestites, intersex and androgynous people, and cross-dressers. The sample aimed to include primarily people who challenged the traditional notions of gender and transgender activists. Here, only preliminary findings are reported and a small sample of respondents have been quoted. Interviewees usually preferred to be identified by name; pseudonyms will be indicated. I have used the terms "transpeople" to cover both transsexuals and transgender people and "trans" to cover transsexuality and transgender. Transpolitics and transtheory refer to politics and theory pertaining to trans states and transpeople.

The postmodern model of trans appears to be at the cutting edge of transtheory and gender theory in general. Various trans authors have argued in support of the fluid, postmodernist world in a manner similar to Plummer (quoted in Ekins and King, 1996:xvi):

> What seems to be sought is a world of multiple gendered fluidities—a world at home in a postmodern cacophony of multiplicity, pastiche and pluralities that marks the death of the meta-narratives of gender which have dominated the modern world. The claim, as Whittle so precisely puts it in his telling contribution to this book, is to "live outside of gender."

Zach Nataf (1996) suggests that transgender highlights gender and sexual ambiguity and fluidity. Kate Bornstein (1994) describes the multiplicity of gendered possibilities while others see transsexuality as a place outside of duality (Cameron et al., 1996). Research findings illustrated the postmodern modal of trans in a number of ways, from the fluidity of transvestism and drag through to the possibly third sex positions of intersex people, transsexuals in transit, and androgynes.

> You accept yourself as then not totally male or female...in some ways I'm not a man, whatever that makes me, it doesn't make me a woman. I wouldn't like the idea of saying that I'm a woman. I don't think I would go that far, but I would go as far as to say I'm in-between, or I'm neither or both or third sex or something like that. (Simon Dessloch)

Sexual orientation logically becomes problematized if genders are fluid, multiple, or not male or female. Trans people have a variety of sexual orientations, with a few, such as David Harrison (in the *San Francisco Chronicle*, Autumn 1995) identifying as pansexual as opposed to bisexual because bisexual assumes two genders. Trans troubles sexual orientation on the individual level, leading some respondents to identify as several orientations at once at different times. It also destabilizes the categories of heterosexual, gay, lesbian, and bisexual on a social level. As Kate More says, it disrupts the means by which sub/cultural membership and sexual attraction are communicated.

> My hair was fairly thick and long and bushy then, and I went down in Doc Martens, jeans, false boobs, makeup on. All the dykes were fancying me until I opened my mouth, and then they thought, "oh my God it's a bloke, ugh!" (Joanna/Dave, pseudonym)

Some of the respondents thought that gender was completely unnecessary on a social level. This would link with the work of authors such as Bob Connell (1987) and Judith Lorber (1994), who discuss the ending of the binary gender system. It was clear from the findings that gender as a rigid male–female dichotomy was fundamentally disrupted by trans:

> I think if you put the end aims of gay people and transgender people together you don't really have much gender left...if both sides get what they want I think it can only lead to a complete breakdown of gender....I think it is possible to get rid of gender to some extent. Of course it depends on how you define gender. I mean I would tend to define gender in terms of not all the little bits and pieces that make up gender, like active, passive, weak, and strong but in terms of putting them together in one package. You have to have one packet which you were born with and you have to have that package until you die. But I think that could easily be and will be gotten rid of. I think like in the past or maybe even the present it's more like a set menu "A" or a set menu "B" and I see the future more like an à la carte menu and you can make your own choice about what you have for starters, for main course or dessert or whatever, or if you're going to have a dessert you can have your dessert for a starter or starter as a dessert or just three desserts or whatever. (Simon Dessloch)

Transgender and transsexual people who envisaged going beyond the gender binary system to allow for longer-term fluidity, third sex or androgynous identities formed a significant minority within the wider trans communities. Some of the groups, such as the Pride working group, the Gender and Sexuality Alliance, and Transgender London, were involved in activism in support of a pluralist gendered society. The transgender people and lesbian, gay, or bisexual trans-people tended to be more challenging of the gender system.

> What I would really like to do is to get a core group of people who feel the same as me about trying to put forward theories about having a society without gender. (Christie Elan-Cane)

The use of feminism by transpeople and as a means of informing transtheory was explored. Any form of feminism is problematic as a basis for analyzing trans in that its locus rests on male-female categorization. The traditional feminist analysis of trans, where transsexuals are seen as reinforcing stereotypes and appropriating female bodies and space, was strongly criticized by participants. Feminism, where it is based on a distinction between women and men, was seen by many contributors to the research and by authors such as Kate Bornstein (1994) as having an investment in the gender system. However, pluralist forms of feminism are more useful to trans politics. Feminists such as bell hooks (1984, in Judith Lorber, 1994) have argued that feminist research based on the male-female binary is flawed; race and class produce many categories. This could be extended to trans and my findings have supported this. For example, Zach Nataf spoke of the similarities between the breakdown of rigid racial categories and trans. Trans is different in that it concerns physiological sex embodiment, but it was theorized by several respondents as forming part of a continuum of gender possibilities.

Poststructuralist feminism usefully informs the understanding of trans on a theoretical level. Cultural feminists such as Judith Butler (1997), Donna Haraway (1991) and Marjorie Garber (1992) challenge the binary opposition and see sex/gender as fluid (Judith Lorber, 1994). Here the body is seen as constructed by discourses of gender that naturalize the essentialist concept of sex. Transsexual practices are seen as the instrumentation by which the body is discursively produced (Stryker, 1998). The transsexual Sandy Stone (in Epstein and Straub, 1992) describes transsexuality as a genre and suggests that bodies act as screens on which academic and medical struggles are projected. Sandy Stone discusses sex changing as performativity, extending the work of poststructuralist feminists such as Judith Butler. A minority of participants discussed trans and trans activism specifically in relation to performativity. Attempting to pass as androgynous, for example, could be seen as a means of challenging gender stereotypes.

Another area of theory that could be used to understand trans is queer theory, in the sense that queer involves the scrambling of binaries and gender as performative rather than innate. As Stephen Whittle (1996:196–214) says: "Queer theorists are attempting to undermine the very foundations of modernist thought: the binary codification of our apparent existence."

However Ken Plummer (1996) usefully points out that queer theory harks back to the binaries that it seeks to transcend. This is not the only criticism of queer, with some feminists seeing it as vanguardist and elitist. Various criticisms of queer emerged from my research, in particular the problems with queer as transgression of social norms concerning gender. Transsexuals have experienced a great deal of social exclusion and many want to fit into the mainstream as a result. One person I interviewed, who had been born intersex, could not cope with the term "queer" because she had been labeled as a freak and systematically excluded from society. Others thought that "queer" was useful because it provided social space for people with nonconventional gender or sexual orientations, a source of pride in being different, and a means of social change based on lesbigay and trans alliances.

There are a number of problems with postmodernist and poststructuralist approaches to theorizing trans. The two main areas emerging from my research were a critique of the decentering of the subject found in the work of poststructuralists such as Judith Butler and a tendency to insufficiently address the institutional and ideological forces structuring gender. Judith Butler (1997) sees agency and identity as formed by internalized discourses. An autonomous sense of self is constructed and lacks any essential base. However, transsexuals experience a sense of self that is different from their bodies. Several contributors to the research suggested that transsexuality was related to innate gender and not simply performativity. For transsexuals this innate gender was felt to be opposite-from-birth-body sexed. However, some transpeople experienced an innate sense of self that was not traditionally gendered. For example, Loren, a female to male [FtM], does not wear a phallus because that would not be "himself" (Whittle, 1996:198). In other instances participants described a sense of self that was beyond both the body and the gender. For transsexuals and some other transpeople the transgendered self was experienced in terms of a strong urge to change. Most transsexuals and some transgenderists feel that trans is not a choice but a necessity:

> I felt impelled to do the thing that I did and it was like a journey, but I didn't have a clue why I was doing it; everything just came into place once I had had the surgery in two stages and done what I had set out to do. Before then I didn't have a clue what I was doing it for; I just knew that I had to do it...I suppose a compulsion was the wrong words, but it was something that I knew I had to do because I could never be at peace with the body that I had had until I had had it done. But I didn't know why. (Christie Elan-Cane)

The trans sense of self cannot be adequately explained by Judith Butler's analysis of the self. To argue that experience of an essential self is simply internalized discourse is to risk accusing others of false consciousness. Moreover, the discrepancies between people's experiences and the discourses surrounding them render a totally constructionist account untenable.

In addition to an identity-based critique of postmodernist models of trans, the postmodern account of gender fluidity, plurality, and performativity can be criticized as lacking structural analysis. My respondents documented numerous ways in which their gender transgression was socially controlled, from parental control through violence at school and on the street to exclusion from the labor market and legal sanctions such as the illegality of marriage.

> well obviously if you were going to be gender-ambiguous then you'd have real problems using public toilets, and even just interacting with people becomes very difficult, and not even mentioning things like being functional in society, like having a job or something like that....a lot of people won't do things because they say, "well I might lose my job, I might lose my family, my relationship, I don't have the money to pay for this and that" and for that reason don't do a thing: I think it depends on as usual who we know and stuff like that. I think if we have certain kinds of connections or certain kinds of qualifications, or experience or knowledge or something like that, you can get around. But if you don't then you're basically just kind of excluded in a way. And if you can be kind of made to fit in, but the question then is how that affects you as a person, if you have to fit into something that you just can't fit in and won't fit in. (Simon Dachsler)

Many transsexual people are forced into potentially abusive situations because of the levels of social control imposed on them; for example, many raise money for their sex change via prostitution, and transsexuals are sent to the birth-sex-specific prison which means that FtMs almost automatically become rape victims.

According to some of the participants, trans is becoming more socially acceptable particularly where it is seen as recreational rather than as a more serious identity change. However, even this depends heavily upon context and is constrained by homophobia as well as transphobia:

> It's Andy's thirtieth birthday party...and he said "you've gotta come." I've got the letter here, I'll read it to you. It was addressed to Joanna. It says: "If you come down dress as you come, as Dave, straight boys and all that." I mean it's a case of "yes." I mean I know I'll go down there and I'll hate it...it's like "what's the point?" It's boring. When I go down there dressed up in drag it'll be fab, it always is. (Joanna/Dave)

The pathologization of transsexuality in particular by the psychiatric establishment has controlled access to treatment and impacted heavily on the gender identities of transsexuals who have been through treatment (see, for example, Stephen Whittle, 1996). In this respect Janice Raymond (1979/1984) was seen to be correct by some participants. Transsexuals are still to an extent forced to adopt stereotypical identities during the period of their treatment. Intersex babies are surgically assigned to one sex or the other at birth, which in the case of one participant has led to serious social, emotional, and physical problems. Overall, the pathologization of trans has led to the silencing of transpeople and to transphobia at both structural and individual levels. Transphobia was seen by participants to have a number of causes. A few of the participants discussed how they personally found trans frightening, particularly the fluid and multiple gender states. Several discussed the way that non-transpeople simply could not cope with the complexity of trans and the way in which it problematizes male and female identities. People felt that transphobia was the result of fear: trans is perceived as threatening. It highlights the extent to which all our identities are constructed and therefore changeable. As Linnae Due (1995) says, "altering our gender rips away the blinders that hide how little of us are ever seen; and how little we see. It's a moment of clarity most of us would rather forget."

Participants identified patriarchy and the Judeo-Christian religion as the origins of transphobia, as cross-culturally and trans-historically trans has been accepted and in some cases highly valued by various other societies (see, for example, Leslie Feinberg, 1996, Kate Bornstein, 1994). Some people discussed how trans was threatening to patriarchy: for example, Alex Whinnom noted the possibility of transsexuality disrupting inheritance laws, which are still gendered among the British aristocracy. The roots of transphobia in religion were seen as various, ranging from the Jewish condemnation of intersex as an abomination to homophobia and phobia concerning non-procreative sexuality. Despite secularization, traditional religious attitudes still permeate society, impacting on the lives on transpeople in complex ways.

The social structuring of trans quite clearly affects the levels of fluidity and the gender permutations that are possible. Postmodernism fails to provide an adequate account of the structural forces affecting gender. Kate Bornstein (1994) argues that transpeople will not attack the gender system until they are free from the need to participate in it. Yet it is questionable how possible it is for any people to exclude themselves from the gender system.

To conclude, an analysis utilizing both postmodernism and structuralism is necessary for understanding transgender and transsexuality. The processes of structuring and fluidity/change on both individual and social levels are concurrent and multifaceted. Some transgender people see gender categories as being temporarily useful as a basis for identity and political action (for example, Bornstein, 1994) and this is supported by the work of feminists such as Gayle Rubin (quoted in Nataf, 1996). Others experience, and seek to make socially concrete, set gender identities. My findings indicated that the usefulness of categorization depends on context. This multiple, context-based picture would lend itself well to postmodernist analysis but could equally well be theorized using feminist standpoint theory (see, for example, Linda Alcoff, 1995). The context-based model is particularly important for understanding the debates about trans as replicating or challenging heteropatriarchy. In true postmodern paradox, trans both challenges gender and accentuates the significance of categories (Bromley, 1995). As Loren Cameron et al. (1996) argue, trans per se does not free us from gender or necessarily perpetuate stereotyping; it simply makes the conventions of gender more visible. This can be seen as an opportunity for learning about the construction of gender, both for transpeople and for wider society.

The fluid possibilities posed by trans call for structuring based on the principles of equality, diversity, and the right to self-determination. Equal rights for transpeople would necessitate an expansion of possibilities for all people and social changes on a structural level to support this, including changes in education, the legal system, and the medical establishment. Transpeople who contributed to my research wished, primarily, simply for the right to live as equal human beings. Areas highlighted for change were official documentation requiring male/female determination, marriage, public toilets, employment policies, the prison service, and the legal, education, and medical systems. However, the complexification and diversification these changes would entail at social-structural levels would not necessarily mean threats, either to people with conventional gender identities or to the institutions of the family and heterosexuality. Transpeople and non-transpeople would have the right to determine their gender and sexual orientation, regardless of whether this was traditional or unusual. Education concerning increased gender possibilities is particularly necessary, given the irrational but understandable fears that many people have concerning trans.

References

Alcoff, Linda. (1995). Cultural feminism versus post-structuralism: The identity crisis in feminist theory. In N. Tuana and R. Tong, *Feminism and philosophy: Essential readings in theory, reinterpretation and application*, Boulder, CO: Westview Press.

Bornstein, Kate. (1994). *Gender outlaw: On men, women and the rest of us.* New York, London: Routledge.

Bromley, Simon. (1995). Border skirmishes: A meditation on gender, new technologies, and the persistence of structure. CRICT workshop: The subject(s) of technology: Feminism, Constructivism and Identity. Brunel University, Uxbridge, U.K., 22–23 June.

Butler, Judith. (1997). *Excitable speech: A politics of the performative.* New York: Routledge

Cameron, Loren, Barbara de Genevieve, and Susan Stryker. (1996). Transpositions: An introduction to Loren Cameron's photography and the art of transsexuality. Unpublished manuscript.
Connell, Robert. (1987). *Gender and power.* Cambridge: Polity Press.
Due, Linnae. (1995). Genderation X. *SF Weekly,* San Francisco, October 25–31.
Epstein, J., and K. Straub (eds.). (1992). *Body guards: The cultural politics of gender ambiguity.* New York and London: Routledge.
Feinberg, Leslie. (1996). *Transgender warriors: Making history from Joan of Arc to Dennis Rodman.* Boston, MA: Beacon Press.
Garber, Marjorie. (1992). *Vested interests: Cross-dressing and cultural anxiety.* New York and London: Routledge.
Haraway, Donna. (1991). *Simians, cyborgs and women: The reinvention of nature.* New York and London: Routledge
hooks, bell. (1984). *Feminist theory from margin to center* Boston, MA: South End Press.
Lorber, Judith. (1994). *Paradoxes of gender.* Newhaven, CT, and London: Yale University Press.
Nataf, Zach. (1996). *Lesbians talk transgender.* London: Scarlett Press.
Plummer, Ken. (1975). *Sexual stigma.* London: Routledge and Kegan Paul.
Plummer, Ken. (1996). Foreword: Genders in question. In Richard Ekins and Dave King (eds.), *Blending genders: Social aspects of cross-dressing and sex-changing* London and New York: Routledge, xiii –xvii.
Raymond, Janice. (1979/1984). *The transsexual empire: The making of the she-male.* New York and London: Teachers College Press.
Stone, Sandy. (1992). The Empire strikes back: A posttranssexual manifesto. In J. Epstein and K. Straub (eds.), *Body guards: The cultural politics of gender ambiguity.* New York and London: Routledge.
Stryker, Susan. (ed.). (1998). *The transgender issue.* Durham, NC: Duke University Press.
Walby, Sylvia. (1992). Post-postmodernism? Theorising social complexity. In Michelle Barrett, A. Phillips *Destabilising theory: Contemporary feminist debates.* Cambridge, England: Polity Press.
Whittle, Stephen. (1996). Gender fucking or fucking gender? Current contributions to theories of gender blending. In Richard Ekins and Dave King (eds.), *Blending genders; Social aspects of cross-dressing and sex-changing.* London and New York: Routledge, 196–214.

Note

The author wishes to acknowledge and thank all the transsexual and transgender people who have so generously contributed to her research.

3. TOWARD THEORIES OF ANDROGYNY

> I grew up thinking that the hatred I faced because of my gender expression was simply a by-product of my nature and that it must be my fault that I was a target for such outrage. I don't want any young person to ever believe that's true again...Today, a great deal of "gender theory" is abstracted from human experience. But if theory is not the crystallized resin of experience, it ceases to be a guide to action.
> —Leslie Feinberg, *Transgender Warriors*

Gender theories in the past have tended toward naturalistic essentialism, namely, that people are born either male or female and that others are freaks of nature, or toward sociological essentialism, a sort of historical determinism that infers from the fact that we are born in our history that we cannot change it through reflection and self-awareness. The chapters in this section contest these theories and try to find alternative ways of relating sex and gender to our experienced truths, often through language.

Lee Anderson Brown, for instance, argues there is something about the binary embedded so deeply in our language, particularly in our pronouns, that we will resist becoming an "it" and slide more easily into acceptance of a "he" who becomes a "she" or vice versa. With the intersex this is not so easy. Our language constrains what we can see and how we can behave. Like Christie Elan-Cane, Brown concludes that changes in our conceptual apparatus through the acceptance of neutral pronouns such as "zie" and "hir" may change not only transgendered discourse but those social practices that lead doctors, intersex persons and transgendered persons to seek normalizing surgery and hetero-normativity.

Avoiding the essentialisms does not remove the need for some theoretical structure that can bypass foundations of physical fact and cultural relativism. The gender fluidity of Judith Butler's poststructuralist performativities is not adequate to account for the feeling of "true" gender identity many transsexuals experience as innate.

Sam Dylan More proposes that our residual certainty of gender identity may be similar to our encoding of native language, where we learn communication codes through immersion in a local context, which picks up secondary and primary language cues. This does not make the recognition of diversity unproblematic, as there is much idiom-stretching to be continuously negotiated. Both gender idioms and those of one's native language create a cross-cultural transgender continuum, making transsexuality a normal variance of human expression, like a foreign language.

More, like Surya Monro, searches for something that can give gender identity some structure or form beyond anarchy or mere whim. In doing this, these authors are raising many other questions arising from a postmodern awareness about the relations between language and what we see as real.

Ashley Tauchert, like Brown, makes the radical move of trying to move gender language beyond Saussurean binary oppositions. She proposes a fuzzification of gender that places androgyny at the center of a broad continuum, demonstrating that there is a middle way between oppressive gender reification and the free-play of the subjective signifier. Her "radical center" offers an agency that can answer the accusations of biological essentialism or naive universalism, and provide an account of "gender" that is neither binary nor chaotic but consistent with the pragmatic Wittgensteinian approach offered in the "Introduction."

Gender as (Native) Language

Sam Dylan More

The question of how a gender identity develops, by what boundaries gender may be confirmed or constricted, and whether gender is an attribute, an activity, a performance, or a language transmitted by the body touches many practical and legal aspects of everyday life. The classification of human behavior as gender-conforming or non-gender-conforming embeds a moral judgment no matter what epistemological model is used.

The essentialist approach that dominates the biomedical scientific world and provides the basis of the legal definition of sex for most countries (More, 1998a; McMullen and Whittle, 1996; Weize and Osburg, 1996) begins from differing procreative capacity and presumes natural causal links between gender, sexual orientation (Behm, 1996), and overt biological sex differences. It causally links gender identity to procreative abilities, sex to genitals and secondary sex characteristics. The guidelines of the Harry Benjamin Foundation and the *DSM-IV* assist in pathologizing transgressions of the essentialist "truths" and establish a pseudoscientific basis for curing or treating any people who do not fit their norm. The numerous exceptions to this "natural" alignment of sexually differentiated anatomy, procreative ability, gender identity, and gender-conforming behavior are generally combined under the derogatory label of "queer" or "other." Some of those whose gender identity or sexual preference are not necessarily connected to their procreative function include transsexuals or cross-dressers, intersexuals, men who are sterile or impotent,[1] prepubescent children, postmenopausal women,[2] and homosexuals.

The essentialist approach has been criticized for naturalizing a social construction. Some argue (Wilchins, 1997:59–62) that simplistic dichotomies like male/female, man/woman, and masculine/feminine benefit a presumptive and compulsive patriarchal and heterosexual economy. For a social constructivist, the body simply inscribes the social text (Butler, 1993). Groups such as Transsexual

Menace subvert essentialist patriarchal and heterosexual values, praising individuals who choose to cross or defy gender barriers.

The debate between essentialists and social constructionists continues, including debate over their different valorizing of "deviant" behavior as constructive or destructive change. As most men or women, however, do exhibit some alignment of gender, sexual orientation, anatomic features, and self-identification, we need to question further whether this is a mere coincidence, a statistical correlation, or naturally caused.

Transsexuals have been scorned by both essentialists and constructivists. While Weize and Osburg classified gender dysphoria as "probably the most severe mental disorder of the human soul," Butler's (1991) support for the performativities of transsexuals is limited as it criticizes the goal of "passing" transsexuals as "dangerous and...stabilizing patriarchy" supporting the very system that has pathologized them so cruelly. Transsexuals, however, provide a useful model to enable us to slide between essentialist and constructivist theory because they break the assumed causal nexus between naturalized body and gender identity. This paper proposes that gender be treated as a naturally acquired native language, social in that it can change over time, but acquired without deliberate instruction or choice.

In this chapter gender will be defined axiomatically as a language, a specific communication pattern that evolves when people interact. This makes gender the medium of an intrinsically collaborative activity. Gender identity may be regarded as a preference of the individual for a specific communication pattern, as different gender patterns emerge when people with different gender identities interact, a preference that is analogical to the deep-rooted desire of most people to express themselves primarily in their own native tongue.

Wickelgren (1976) states that "The ultimate goal of speech is to communicate meaning from one person to the other." Language as a vehicle for meaning, especially in its most common form of face-to-face conversation, not only involves verbal, semantic activity but relies on a highly distinctive nonverbal code (Laver, 1976). The nonverbal contents of language (Abercombie, 1967; Maccoby, 1998) include body language, posture, proximity, gesture, facial expression, movements and in a wider sense also clothing style, haircut, and accessories like jewelry. Voice modulations like pitch or stress sit between verbal and nonverbal language and provide one of the most powerful vehicles for the expression of emotion (Crystal, 1969). Applying Cicourel's and Abercombie's model of indexing, we will analyze for their interpersonal and sociopolitical significance the communication patterns that evolve when transgender elements are introduced in conversation.

To communicate successfully, speakers first have to agree on a suitable code. Let us consider a hypothetical pair, Al and Bill. Apart from agreeing on a mutually accessible language code like English, other appropriate sub-codes have to be selected. This is usually negotiated nonverbally by that pair checking each other out and modifying the individually selected code by making assumptions about the gender, social status, age, professional experience, and emotional closeness of the other. This process has been labeled indexing. This first (primary) indexing provides the basis against which further information will be checked for reliability during the conversation.

Only the second step introduces verbal information. Al: "Hi, how are you? Glad to meet you, finally." Bill registers this information, but at the same time a second indexing step is introduced. He checks to see how true and reliable this information is. This should not be regarded as malicious distrust, but as a salient factor of everyday communication. People check for irony, for jests as rhetorical patterns, but also for non-reliability in a narrower sense: erogenous information and deception. In order to accomplish this they constantly check the consistency of information and compare primary indexing markers to estimate concurrent secondary body language in order to investigate how reliable the information might be. If a discrepancy between primary and secondary indexing markers, which were responsible for the choice of the gender-language code, occurs, Bill will infer that the true meaning is not that Al is "Glad to meet [him] finally" but something else: a jest or even a hostile ironic statement.

Bill replies, maybe equally harmlessly, "Thanks Al, fine. I hope you haven't been waiting too long?" and Al will proceed with his turn of indexing, checking and so on. As people commonly exchange more than a few sentences, minor deviations from the norm do not prevent communication. As long as the overall pattern is consistent, the communication can be rated as successful.

In normal communication, indexing markers, content, and code are in more or less harmonious alignment. Usually gender, age, social status, and professional experience are correctly assigned and a pleasant conversational atmosphere is established. Some of the indexing markers are common to most human cultures but some are not. What is close to an Arab is intrusive to a North American (Hall, 1963); Japanese dictionaries explicitly list and explain Western facial expressions or customs that are not found in Japanese culture. Many are gender-specific, justifying the axiomatic assumption of this chapter that different codes of gender operate as a language code, attributing arbitrary meaning to certain elements and using them to communicate. Accordingly, two codes may be perceived as sufficiently separate when they attribute different meanings to the same or closely related "word" or syntax combinations. In this chapter, communication will be rated as successful when the original meaning is transmitted and recognized by the receptor of the communication.

Usually the selected gender code crucially depends on the partnership involved in the communication. There is no male or female pattern per se, but male/male, female/female, female/male, and male/female patterns, especially marked in the Japanese language. People usually possess the ability to communicate in more than one of these codes. Having defined gender identity in this chapter as a deep-rooted, unchangeable preference for a communication pattern, we will assume further that most people can only fully communicate in two sets of codes: female/female and female/male or male/male and male/female, forming the female and male gender identity respectively. Gender identity will then correspond to the linguistic term of language competence for a native language, while gender performance will parallel language performance and become a cooperative, self-organizing structural medium for communication especially in conversational interchange.

This is consistent with recent experimental studies (Maccoby, 1998), which have demonstrated that gender differences are shaped by the respective interactions and reactions of the interlocutors. One of the prominent gender-specific factors in face-to-face conversation, for instance, is pitch and loudness of tone, resulting in a different interpretation of speech modulation for men and women (Laver, 1976). A normal male speech modulation will accordingly be perceived as aggressive when it comes from a person who is registered as female. The necessity of distinguishing between intergender and intragender communication can be illustrated by the length of eye contact. While lengthy eye contact is regarded as flirtation between men and women, the same length is usually interpreted as a sign of tension or aggression when it occurs in a female/female or male/male dialogue. Female group-integrating behavior is read among men as submissive. Similar degrees of assertivity are interpreted as aggressive behavior in women and acceptable among men. As with proximity, culture shapes the permissible distance or length of eye contact (Hall,

1963), introducing apparent cross- or transgender elements into conversation when people from different cultures interact

Apparent similarities between the male and female code may in fact disguise separate codes. Exclusively verbal gender-specific expressions exist: a pretty and competent young woman may be described in both German and Japanese as "clean" (*sauber* in German, *kirei* in Japanese), but applying the same attribute to a man would rather hint at unreliability and lack of responsibility (German) or indicate homosexuality (Japanese, female code). This contradicts the theory that only one code exists and that male and female sub-codes differ only in frequency due to a dominant patriarchy (Maccoby, 1998). It also suggests the risk of an eventual misinterpretation due to a code falsely assigned during the primary indexing phase.

Further, words borrowed from other language-systems often shift in meaning as they cross codes; for example, the German expression *Arbeit* (work, profession) is the origin of the Japanese *arbeito* (part-time work; in German *Aushilfs*). Native Japanese speakers, however, interpret *arbeito* as *Arbeit* rather than *Aushilfs*. The difficulty of communicating across cultures is in this way similar to transgender communication in revealing the subtle codes we have learned unconsciously in our native tongue.

Because these paralingual features are culture-specific, Japanese and Europeans may often regard their respective counterparts as "over-polite" or rude (Torrai, 1998). Many traditional Asian customs like women opening doors for men, or men helping someone into a coat, assign gender roles contrary to those in Western society. An analysis of these and other less prominent transgender elements might facilitate a better understanding between members of different cultures. Because trust is a necessary condition of communication, internally inconsistent messages are received with caution. The style of the conversation is controlled by two conflicting individual motives: the need to clarify one's position and the need to protect formerly established norms and habits.

What happens when transgender elements are introduced into the communication? Transgender individuals, whether the sissy boy, the masculine woman, the butch dyke, the tomboy, the drag queen, the post-operative transsexual women in jeans with the large hands and the deep voice, and/or the boyish-appearing intersexual, share in all their diversity one prominent characteristic: they exhibit features that conflict with the primary indexing that has led the partner to assume a communication code and that results in giving the secondary indexing markers a non-indented meaning. This can interfere with the primary indexing and lead the communication partner to choose the wrong code: for example, female/female instead of female/male. This situation defines the transsexual case.

Another option is that although the communication partner picks a code that the transgender partner would agree with, various incompatibilities with normed gender responses might induce consciously or subconsciously recorded code switches. An example of this might be a transgender butch lesbian, who prefers several but not all masculine gender expressions. Other examples might be provided by those who would never identify consciously as transgender: individuals from a different cultural background, who have learned different gender expressions, which now create a similar pattern of mixed messages presenting toward the non-transsexual observer and communication partner conflicting communication strategies.

In both cases, primary and secondary indexing will be in permanent conflict, resulting in possible misunderstanding of the intentions of the transgender person. As this subconscious indexing is based on socially accepted and commonly learned communication patterns between adults, it is not questioned by the interlocutor. Similar communication miscoding occurs when people with laryngitis who can only whisper are generally also answered in hushed tones, the whisper being interpreted as a sign of personal intimacy and not correctly identified as due to a medical condition (Hall, 1964).

When communication is dominated by language patterns that vary from the established pattern of the dominant culture or merge the binary gender language codes, it is prone to misinterpretation. Communications are read as blurry, as "mixed messages," or as hesitation or an inability to make a clear-cut decision. Subsequently the non-transgender communication partner may become confused, irritated, or even offended. Daily experiences help us to infer that where the content and body language of individuals are in constant mismatch and there are no indicators of irony or jest, the inconsistent individual is not revealing his true intentions. As most people are not transgender, this habit is difficult to override and may even lead to a further destructive pattern: the partner of the transsexual may not seek further clarification as this could reveal his own weakness toward a potentially dangerous person.

More extreme reactions might involve emotional withdrawal, assertion of status, intimidation, or even physical attack or rape. It has been shown that approximately 50% of transgender, transsexual, or intersex individuals experience domestic violence or are raped by a romantic partner (Lev and Lev, 1999), an indicator of dysfunctional communication. Other figures indicate that gays and lesbians experience a higher occurrence of parental abuse, taunts, or harassment in youth and adolescence, which cannot be attributed to actual sexual behavior, suggesting that transgender elements need not be linked with a conscious transgender identification to be severely punished.

Many transgendered individuals try to change the code by emphasizing accessible markers in an attempt to subvert the primary indexing and to enforce a code switch. Unfortunately, first impressions in conversation are difficult to override so this behavior is commonly interpreted as "overcompensation." Transsexuals have been frequently rated as hyper-feminine or hyper-masculine by critical voices of the queer and the medical community. Their transgendered communication pattern is regarded as dishonest, and the cycle of distrust and miscommunication continues.

In the first case, where only the primary indexing has led the communication partner to assume the wrong code, both parties would benefit from a code-switch, which would convert the previously dysfunctional communication into a successful one. As successful communication is essentially an interactive and consensual performance activity between people, a refusal of the appropriate code severely limits the options for self-expression and communication of the person who is denied communication appropriate to his or her language competence (gender identity). Disrespecting a gender identity might be accordingly regarded as a kind of forced cooperation and abuse.

As dominant Western epistemologies do not admit nonalignment of either gender identity and body (essentialist) or socialization and body (constructionist), many transgendered individuals see changing their body morphology, for example, by sex reassignment surgery or transition, as the only way to mutual acknowledgment and acceptance. This requirement, however, leads to further complications.

As the mutual aim of conversation is to assess each participant's identity, any "gender switch" or "sex change" threatens reliability. An attempt by a transsexual to clarify the primary indexing process by amending its physical markers may be perceived as a danger to the continuity of the self. In transition, both communication partners must accept mutual responsibility for creating successful new communication codes. While negotiations about the appropriate gender can indeed present an alternative solution, body-transformation is time-consuming and requiring the consent of the communication partner, calls for a more intensive relationship. As passing obviates this negotiation phase, many transsexuals choose to alter markers that have led to an erroneous primary indexing, preferring to cut their ties with previous gender markers. The benefit of transition has been documented as a dramatic improvement in the quality of life.

Perceiving gender as a language resulting from a collaborative performance activity, rather than a natural attribute owned by or assigned to an individual, predicts this improvement in life quality, as surgery and/or hormonal treatment facilitate the transmission of normal codes of gender identity. This liberation also takes place with no outer audience present, as self-reflection is also a kind of communication, the special case in which transmitter and receiver are the same person. When one is not forced to prove specific gender identity all the time, one gains extra energy and inner peace (More, 1998b).

The linguistic model presented here also calls for a critical re-evaluation of current laws that, using essentialist criteria, require a "complete" physiological transition instead of a coherent external one, which would be sufficient to allow a corrected primary indexing mechanism. Transmission of informational content requires only consistency of expressed language elements but not the inclusion of all possible language elements. For the transgender/transsexual case this would imply that extreme sex reassignment surgery would not be required for a change of legal status. This also affects the present standards of care, which require transsexuals to "pass" as members of the opposite sex for a lengthy period of time before surgery is allowed, often with major discomfort and dissociation from their own body. The requirement of despising their own bodies as a diagnostic criterion for transsexualism (DSM-IV) also affects intersexuals as it pathologizes their natural bodies, denying these groups pride and erotic pleasure in their own physical embodiment.

For example a female to male [FtM] might not need sterilization, or a penis to qualify as a man. Surgery should be an option for those individuals who regard it as a necessary ingredient for personal self-expression or as an aesthetic or functional requirement for sexual intercourse. The special case of conversational activity is a private performance activity. In contrast, a public demonstration of "sexual behavior" might induce tension in observers, as a conflict of private/public codes. The implication of sex already implied by the term "transsexual" may have contributed toward the stigmatization of transgender identities.

Murphy-Shigematsu (1997) suggests that essentialist physiological racial classifications based on the "one-drop-of-blood rule" assign Japanese-Americans to an inferior racial category. If we focus on the cultural communication differences in body language and social practices (Ueda, 1996), we do not remove the potential to discriminate negatively, but we allow minority groups to participate more easily in social change and actively challenge ethnic boundaries.

Defining gender as a language might help transsexuals avoid "therapeutic" measures that transform them superficially into users of a commonly accepted code. It might also help gays and lesbians (also gay or lesbian transmen or -women) who do not seek a change of their officially and legally acknowledged "gender identity" but would profit from a more relaxed and less anxious attitude toward code switches in conversation. Interpreting the use of different codes as a deliberate or culturally predisposed affinity for certain communication patterns, this model evades the association with sex and the stigma attached to an obvious or apparent transgression.

Classifying transgender elements in a more neutral light as an introduction of code-switching elements in order to make speech salient to interlocutors and to enrich and vitalize its quality allows a perception of gay and lesbian partnership to focus less on their sexual preference, and evades the mind/body split experienced by transgender individuals (Martin, 1994). Fotos (1995) described a similar process in the effects of English-Japanese code switching in conversation. As gender is defined as the medium for a cooperative exchange activity between persons, the individual's own mind/body integrity is independent of gender. The transgender effect is created by a mismatch of expected/perceived language performance. Because the body is both a physical carrier and an indispensable tool for the transmission of information via the gendered code, any mismatch of gender codes creates the transphobic distrust.

Certain types of attraction among homosexuals[3] could be analogous to the attraction of bilingual speakers who have the undeniable advantage of mutually comprehensible code switches, providing a deeper mutual exchange and option for self-expression. As the benefits of this code-switching ability can only fully be realized with a cooperative partner who has bilingual access to both codes, it emphasizes the inherently consensual aspect and the empowering nature of same-sex relationships, denied by essentialist models.

Judging sexual orientation as well as gender identity as a general communication preference and the ability to communicate in a gender-bilingual context would define gay/lesbian identity as the ability to recognize the gender bilinguality in others without being trapped in the pitfalls of the primary indexing mechanism. Identity transcends the immediate romantic partnership. A femme who dates a passing butch would not be diminished in their lesbian identity. This relationship would demonstrate her (lesbian-transgender or "engendering") ability to cope with and enjoy transgendered elements in communication with her partner. Accordingly, gay people would naturally be attracted to partners who understand gay codes—and not necessarily to partners of similar genital anatomy. Indeed, transsexuals or intersexuals who do not always possess the standard anatomy successfully find gay or lesbian partners who do not identify as bisexual.

Moving beyond essentialism to a treatment of gender as native language might increase the understanding and acceptance of transgendered expression in general as it reveals one of the origins of negative discrimination—the fear of non-transgender or monolinguistic people that transition is a non-stable, or pathological behavior that cannot be trusted. One could analyze gender code switching as objectively as normal bilingual language code switching to reveal ways of emphasis, clarification, getting and holding attention, identifying particular topics, reporting speech, signaling that a repair to a previous utterance would follow, and making part of an utterance prominent and dramatic. Such discourse analysis contrasts with the pathologizing classification of transgender behavior reflected in both medical literature (Zucker and Bradley, 1995) and legal documents (Whittle, 1996). In the fight of inter-sexuals and transsexuals for equal civil and human rights this approach might lead to the higher self-esteem already demonstrated by the gay and lesbian liberation movement. Perceiving gender from the position of linguistic interpretation may help toward that goal.

References

Abercombie, D. (1967). *Elements of general phonetics*. Edinburgh: Edinburgh University Press.

American Psychiatric Association Task Force. (1994, 4th ed.). *Diagnostic and Statistical Manual of Mental Disorders: DSM-IV*. Washington, DC: American Psychiatric Association.

Arnold, A. P. (1996). Genetically triggered sexual differentiation of brain and behaviour. *Hormones and Behaviour*, 30, 495–505.

Atkinson, Leslie, and Kenneth J. Zucker (eds.). (1997). *Attachment and psychopathology*. New York: Guilford Press.

Behm, D. J. (1996). Exotic becomes erotic: A developmental theory of sexual orientation. *Psychological Review*, 320–335.
Butler, J. (1991). *Gender trouble: Feminism and the subversion of identity.* New York and London: Routledge.
Butler, J. (1993). *Bodies that matter: On the discursive limits of 'sex.'*. New York: Routledge.
Cicourel, A. V. (1972). Basic and normative rules in negotiation of status and role. In D. Sudnow (ed.), *Studies in social interaction.* Cambridge, MA: Harvard University Press.
Crystal, D. (1969). *Prosodic systems and intonation in English.* New York: Cambridge University Press.
Fotos, S. (1995). Japanese-English conversational code-switching in balanced and limited proficiency bilinguals. *Japan Journal of Multilingualism and Multiculturalism*, 1:1, 2–15.
Hall, E. T. (1963). A system for the notation of proxemic behaviour. *American Anthropologist*, 65, 1003–1026.
Hall, E. T. (1964). Silent assumptions in social communication. In D. Mc. Rioch and E. A. Weinstein (eds.), *Disorders of communication*, Baltimore: Association for Research in Nervous and Mental Diseases.
Laver J. (1976). Language and nonverbal communication. *Handbook of Perception, Vol. VII, Language and Speech*. New York: Academic Press.
Lev, Arlene Istar, and Sundance Lev. (1999). Sexual assault in the lesbian, gay, bisexual and transgender communities. In *A professional's guide to understanding gay and lesbian domestic violence: Understanding practice interventions*. Ed. Joan C. McClennen and John Guther, Lewiston, NY: The Edwin Mellen Press.
Maccoby, E. E. (1998). *The paradox of gender—The two sexes: Growing up apart, coming together.* Cambridge, MA: Belknapp Press of Harvard University.
Martin, B. (1994). Sexuality without genders and other queer utopias. *Diacritics*, Summer-Fall, 104.
McMullen, M., and S. Whittle. (1996). *Transvestitism, transsexuality and the law.* London: Gender Trust.
More, S. D. (1998a). The pregnant man: An oxymoron. *Journal of Gender Studies*, December, 5:54.
More, S. D. (1998b). Scars. *FTM Nippon Communications*, December, 4:10.
Murphy-Shigematsu, S. (1997). American-Japanese ethnic identities: Individual assertions and social reflections. *Japan Journal of Multilingualism and Multiculturalism*, 3:1, 23–27.
Torrai, M. (1998). Unpublished letter to the author.
Ueda, Y. (1996). Japanese compliment responses: A comparison to American English norms. *Japan Journal of Multilingualism and Multiculturalism*, 2:1, 52–62.
Wedge, B., and C. Murumcrew. (1965). Psychological factors in Soviet disarmament negotiation. *Journal of Conflict Resolution*, 9.
Weize, C., and S. Osburg. (1996). Transexualism in Germany, empirical data on epidemiology and application of the German Transsexuals Act during its first 10 years. *Archives of Sexual Behaviour*, 25(4), 409–25.

Whittle, S. (1996). Gemeinschaftsfremden—Or how to be shafted by your friends. Press for Change, London. <http://www.pfc.org.uk/legal/whittle4.htm>.

Wickelgren, W. A. (1976). Phonetic coding and serial order. *Handbook of Perception, Vol. VII, Language and Speech.* New York: Academic Press.

Wilchins, Riki. (1997). *Read my lips: Sexual subversion and the end of gender.* Ithaca, NY: Firebrand Books.

Zucker, Kenneth J., and Susan J. Bradley (eds.). (1995) *Gender Identity Disorder and psychosexual problems in children and adolescents.* New York: Guilford Press

Notes

1. Huge sales of Viagra prove the psychological problem this presents to most men.
2. Perhaps the lower social status of women could be due to their entering a non-procreative state earlier than men.
3. Persons who have only proficiency in one unigendered code (female-female, male-male) would naturally be also more attracted and able to form stable relationships with persons who are proficient in that code, thus generating a different, non-transgendered variant of homosexuality.

Beyond the Binary: Fuzzy Gender and the Radical Center

Ashley Tauchert

Within the disciplines of the humanities and cultural studies, gender is understood in the context of "binary oppositions." The "binary" as a cultural and conceptual phenomenon can be illustrated with reference to Hélène Cixous' opening to her essay "Sorties," which has become an invaluable—if problematic—touchstone for explaining the theoretical relationship between "binary oppositions" and gender to students. Cixous (1975:14) lines up binary oppositions in the following way, and argues that all binaries are derived from, and return to, the original pair: male/female:

>Activity/Passivity
>Sun/Moon
>Culture/Nature
>Day/Night
>Father/Mother
>Head/Emotions
>Intelligible/Sensitive
>Logos/Pathos
>Man/Woman

I'll return to this list later, but first I want to address binary structures in general.

Binary opposition is the theoretical model on which language makes its sense: black/white, male/female, left/right, dark/light, night/day and so on. Every signifying term has its opposite, against which it defines itself. The binary as a cultural concept is extrapolated from the work of the Swiss linguist Ferdinand de Saussure,[1] whose investigation of the structural properties of language as a system have deeply influenced modern thought. In Saussurian terms the word "up" cannot function without conjuring and cancelling out the properties of its opposite, "down." Every concept relies on a chain of negated differences to maintain its claim to meaning. We can have no concept of "up" without a concept of "down" to negate. This abstract model, then, can be applied to any cultural language that deploys binary oppositions to make sense of the world. For example, any textual or visual reference to the concept of "man" always necessarily conjures and negates the concept of "woman," and the qualities of "masculinity" are necessarily dependent on the negated presence of "femininity." Similar points can be made about black/white, queer/straight, empowered/disempowered, and so on. This model of language and concepts exposes a tendency to polarize categories in order to separate and clarify our objects and position-ality, and ultimately our identities: oppositionality demands that we identify ourselves with one category or another; there is no orthodox space for between-ness. For those of us in the Humanities interested in investigating the interface between a codified binary cultural model of gender (masculinity and femininity) and the lived experience of the range of embodied subjectivities that are under pressure to conform to this model, whether their embodiment is in accordance with it or not, are engaged in a range of discussions concerning the nature of the binary, which can be characterized as a late twentieth-century version of the "nature/nurture" debate. The conclusions drawn from this theoretical model to date are, however, remarkably limited in their scope. Having noticed the function of binary models in thought and language, we seem only to be able to conceive of the following three general responses:

(1) to invert binary weightings and values (displace currently over-valorized terms with currently under-valorized terms);

(2) to disturb the binary process by introducing a third term, or by circumlocuting the binary terms in language that has become known as "politically correct"; or

(3) to undermine the binary as an illusion imposed on a more general and de-hierarchized web of "difference" between a whole range of embodiments and identities, and replace it with the floating free-play of the signifier that is the hallmark of post-modernism.

In terms of the *gender* binary that we are engaged with today, these three positions translate into the following:

(1) (for example) to valorize the previously repressed and eclipsed feminine at the expense of the historical dominance of the masculine;

(2) to deploy a gender-free language that challenges binary modes of thinking (the best example of this is the "perkin hole"

in place of the "man hole," but I have also tried using "zie" or "hir" in place of he/she and his/her);

(3) to speak of gender as a localized and fallacious symptom of language systems, which is what we find in many postmodern or poststructuralist responses to the question of gender. An extreme —if implicit—case in point is Foucault's gender-blindness throughout his analysis of subjects and power (McNay, 1994).

Interestingly, it is only the last position that raises the intersex subject and/or transgender (in the embodied form of hermaphroditism, or in the codified and very serious play of drag and parody) as an argumentative strategy, and in that scenario, intersex/transgender is offered as that which exceeds and undoes the binary: neither male nor female, and not approximate to either position.

This paper offers a response to the gender binary that accounts for the intersexed subject not as that which exceeds cultural hierarchies, but as a new orthodoxy in itself.[2] The model I am offering I have designated "fuzzy gender" as it draws on work in the fields of mathematics and physics. These fields deploy the term "fuzzy" to denote a non-Aristotelian conception of system, structure, or process; that is, a conception that moves beyond the binary of either/or, without collapsing into the chaos of free-floating subjectivity in difference. I should make two things clear at this point: I am keen to find a way of speaking of gender that does not trap me—as a feminist theorist—in unsustainable and undesirable positions of essentialism or universalism; secondly, I want to foreground the ethical parameters of this chapter as a way of organizing our thinking about gender. Gender is not only a theoretical term, it is also a political term; that is, what we make of gender determines what is possible in the social and material world, for, after all, in the present cultural economy we are all either male or female or excluded from normative humanity altogether. If gender is only a symptom of discourse, the suffering of the subject that does not fit the binary gender model is as real as any suffering, and the pains caused directly and indirectly through the discourses of binary gender are insupportable. Or, as Judith Shapiro (1991:249) puts it, "While transsexualism reveals that a society's gender system is a trick done with mirrors, those mirrors are the walls of our species' very real and only home." But the notion that to "deconstruct" the binary is a move toward free-floating subjectivity offers little scope for political and effective agency. I offer my ethical position first, therefore, and it is simply this: *means justify ends*. I'll return to it, and to the intersexed subject in whose name it is spoken, in a moment.

First I'd like to point to some peculiarities of the binary as described by Cixous.

Binary Opposition and the Continuum Between Opposites

Some of the binaries noted by Cixous in her influential essay are necessary "opposites" in the sense that they cannot be simultaneously present (Day/Night). It is worth bearing in mind, however, that the opposition indicated and captured by the binary is always a fiction: not in the sense of an illusion, but in the sense in which days of the week and months of the year are necessary fictions. Claiming that Day/Night is a "binary opposition" is a complex maneuver in itself. Claiming that Day/Night is a "binary opposition" that is false and needs undoing is perversely complex. But perversity in this discussion is not to be avoided.

Day/Night, then, is an "opposition" insofar as (in Aristotelian terms) *A cannot equal not-A:* Day cannot equal Night; it cannot be *both* Day *and* Night at the same time; it is *either* Day *or* Night at any one time. OK. But there are moments in the Day/Night when it *is* both Day and Night at the same time (conceptualized as Dawn and Dusk—the equinoxes of the Day/Night; the 50% points that approximate neither to day nor to night, but which contain, equally, aspects of both). What Cixous is noting in her list of binaries, then, is not something about the world as such but something about our practice of conceptualizing the world as such, which is closely related to (but not identical with) language. Our concepts of Day and Night as discrete entities—and by implication opposing entities—influence the way we experience the Day/Night. But there is no exact and definite moment at which "it" stops being Day and starts being Night. Day and Night exist on a continuum that can be more accurately conceptualized as a circle. Similarly, the oppositions between winter and summer, dark and light, black and white, old and young, in and out, up and down (and so on) describe given points on a continuum. So that what we are describing when we deploy the term "binary oppositions" is not the tendency of the world to divide itself into oppositional entities, but the tendency of our culture to divide continuums into oppositional and discrete entities. The oppositional insistence seems to result from the act of differentiation itself: in order to maintain difference between points on a continuum we have to draw a boundary. Because the boundary cuts through an ever-changing continuum it has to be very carefully policed (enforced) for the categories to remain differentiated at all. The boundary, or axis, takes its place in order to differentiate (binary oppositions exist because we need to differentiate between objects and concepts). The act of differentiation is reified to become the static concept of binary opposition as a result of the degree of oppositional force necessary to police the boundary of differentiation across a continuum.

I offer this model for thinking through binary oppositions as a contribution toward ongoing arguments against the essentializing of gender as natural oppositional kinds, and in favor of the de-essentializing of gender on the model of a continuum between masculinity and femininity.[3] But for the purposes of this chapter, I want to develop my claim that where we find binary oppositions, we find very carefully policed and enforced boundaries between otherwise disturbingly permeable and fluid categories. The oppositional conception of masculine/feminine that operates in our culture, then, is a useful example of the reification of the oppositional tendency toward static and discrete concepts. That is, the degree to which masculinity and femininity are culturally defined as oppositional—and evidence of this is available in any newspaper on any given day—indicates and emphasizes the degree to which our culture finds it necessary to enforce and police a boundary between points on a continuum. The more absolute the binary concept, the more fluid the boundaries. My argument, then, can be developed along the lines that gender is neither totally a fiction, nor totally essential, but somewhere in between (as in the Day/Night continuum being subject to the cultural fiction of Day or Night).

This is *crucially not the same* as saying that there is no necessary differentiation outside discourse between masculine/feminine, and that any discussion of gender difference is merely an imposition by discourse on a chaotic and undifferentiated material. As in the case of the Day/Night, it is not so much that there is *no difference* between Day and Night, and that the differences we experience are discursively produced. Day is light and Night is dark for a start.

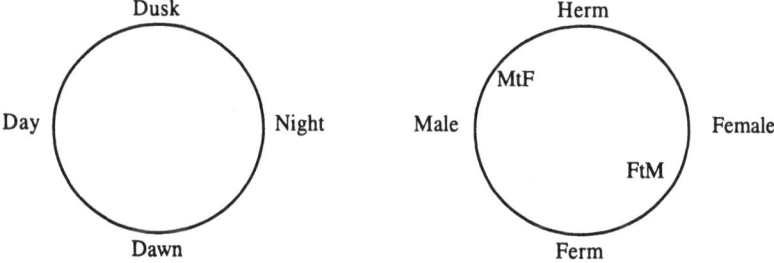

Figure 3. Binary continua

What I am trying to define here is the difference between believing that Day and Night are absolute opposites, and remembering that Day imperceptibly becomes Night, Night imperceptibly becomes Day. The binary is the moment at which we demand that a gradual and imperceptible continuum become a static and oppositionally understood category. I'd like to apply this argument to "gender" more specifically to see where it takes us.

Judith Butler has argued convincingly, in *Gender Trouble* (1990) and *Bodies That Matter* (1993), that the urgency with which we differentiate between sex and gender is a suspect move, and one that allows us to maintain that sex is a natural and unchanging essential truth, while gender is a cultural construction. Butler contests that what we think of as "sex" is in fact *embodied gender*: the corporeal incarnation of a discursively constituted (performative) gender. I follow Butler up to the point where she denies any material force in the constitution of gender, but as a feminist academic activist I see good reason to defend the category of "woman" in theoretical and political work, because without this category there is no feminist political agency, and without feminist political agency, there is no interrogation of the gender power imbalance in our institutions of knowledge for a start—but that is another argument. I tend to use in theoretical work the term "female-embodied subject," by which the "female" designates embodiment that corresponds with the material factors recognized as "female" in a given cultural-historical matrix. I am fully in agreement with positions that display the constitutive power of gender as a discourse, and the materialization of gender *as* anatomical sex. But I fear that by jumping from the static binary immediately into a gender-free modality we are missing something interesting that lies in the middle.

So I am going to move now to the "radical center" between the binaries, between sex and gender (which, along with Butler—but in a slightly differently constituted matrix—I believe to be the same thing), between male/masculine and female/feminine, between the reification of gender into immanence and the a-political and a-ethical soup of utterly deconstructed and de-gendered subjectivity. I'd like to introduce a model of "fuzzy gender" as a way of accounting for embodied subjectivities that do not conform to the binary model that dominates our culture. "Fuzzy" is a term I have borrowed shamelessly, and to the disgust of my mathematical colleagues, from a field of mathematical logic called "fuzzy logic."[4] My familiarity with "fuzzy logic" comes from the popular work of Bart Kosko (1993), and I apologize in advance to any mathematicians or physicists for crudifying what I appreciate is a very specialized area of research. "Fuzzy logic" in mathematical terms is a theoretical model that accounts for the inevitable ambiguities that occur to trouble any precise mathematical formula. In binary Aristotelian logic, the law of non-contradiction reigns (A cannot = not A); something is either A or not A, and anything that doesn't conform to this model is categorized as an ambiguity that—it is assumed—will be disambiguated by the accumulation of more detailed knowledge about the thing under definition. Fuzziness allows for ambiguity, not as a problem to be resolved, but as the "excluded middle" that is simply ignored in Aristotelian logic. For example, a platypus is both a mammal and not a mammal. Aristotelian logic usually accounts for such anomalies with recourse to probability theory, which calculates the probability that a platypus is a mammal. In Aristotelian logic, all things are either black or white. In Fuzzy logic, all things are black and white (gray) to some degree, but at the poles of the spectrum there are special (extreme and polarized) conditions of grayness. Aristotelian logic only recognizes these special (extreme/polarized) cases as valid categories, and rounds up a point on the spectrum to either approximately white or approximately black. The problem for Aristotelian logic lies with the middle of the spectrum—the 50% point—where a thing is both black and white to the same degree, and cannot be approximated to either pole, for example, the platypus, or the half-full/half-empty glass of water, which is full and empty to the same degree, or the intersexed subject. In fuzzy logic, the 0/1 binary is posited as a continuum, so that an infinitude of points are recognized *between* the polar extremes of 0 and 1, and the 50% point that approximates neither to 0 nor 1 is recognized as such.

If we apply this model to gender, something interesting happens. In the dominant cultural model of binary gender we are all *either* male *or* female, and must approximate to one position or the other. Ambiguous cases are designated as illness, monstrosity, or the result of unfortunate birth defects. Truly ambiguous cases, such as intersex, are hushed up and surgically modified to fit the culturally recognized model of male or female. In fuzzy terms, the culturally dominant model only recognizes the polarized extremes of a continuum of infinite points: on a model of fuzzy gender, we are all male *and* female *to some degree* (fuzzy math uses the language of percentages to describe degrees), and the 50% point (intersex) is recognized as a special case that is both male *and* female *to equal degrees.* The dominant cultural model of binary gender, then, valorizes the extreme polarities of a continuum, and pressurizes or disciplines all embodied subjects to conform to the peculiar properties of the binary poles of that continuum, in a similar way to the Day/Night example I discussed earlier.

My point, then, is that it is the act of differentiation—when insisted upon against a counter force towards continuum—that results in binary oppositionality. Another way of thinking this is to say that binary opposition is a verb rather than a noun. Those qualities that are differentiated by culture become reified as opposites (up/down, in/out, black/white, you/me, subject/object, man/woman, masculine/feminine). Conceived (conceptualized) as continuums rather than opposites (everything is relative after all, as the Grand Old Duke of York realized in the end), the act of differentiation is a matter of perspective. Perspective relies on positionality. And positionality can be understood through perspective. Whether an understairs cupboard door opens inwards or outwards depends on the point of view of the trapped cat. That gender is *neither* a fixed binary *nor* a chaotic anarchy can be evidenced by thinking about the instances that exceed the binary cultural model: transgender, bisexuality, transexualism, intersex, Androgen Insensitivity Syndrome, queer sexualities, facial hair on women, Turner's Syndrome, hypogonadotropic hypogonadism, hairy legs on women, and so on. All of these—and countless more—embodiments have been pathologized to varying degrees by the binary economy that governs gender identity. Under this binary economy that governs our cultural perception and organization (including identity), these conditions are characterized as mistakes, failures, monstrosities, illnesses, unfortunate cases; whether in a repressive cultural régime that demands conformity to the binary normative human subject (male or female in a strictly Aristotelian sense), or whether in a more liberal régime that allows subjects that exceed or recede from the binary model space to exist, but as a gift, in a gesture that perpetuates their exclusion from normative human subjectivity (for example, the global debate about whether transsex operations should be funded by the national health system).

Intersex is designated the most monstrous of all forms under the present binary economy, because it raises a question about the nature of morphological gender. It is current practice to intervene surgically to adapt these corporeal subjects by surgery as soon as possible after birth, in order to approximate them to one of the given binary genders; male or female. Although the most recent statistics suggest that between 1% and 4% of the population is born intersex in every generation, this embodiment has been pathologized to such a degree that it is invisible and unthinkable. Our cultural binary model depends on the view that although gender is deemed to be natural, nature is also deemed to make mistakes that culture can correct. This seems to me to be a case of having and eating the same cake. If we assume for the moment that "nature" does not make mistakes, but that embodied subjects materialize in a whole range of sexual embodiments—from 100% male-embodied to 100% female-embodied—we can apply our model of the radical center to a non-binary gender and see what happens.

In this (imaginary) model, then, the cultural prototype to which embodied exemplars approximate is not binary, but neither is it chaotic; it is *singular*. It is a model that takes the intersexed subject as the normative human subject, to which all embodied subjects conform *to some degree*. Applying this model would engender a cultural model of embodied subjectivity that excludes *no* embodied subject from cultural normativity; that is, the model allows that nature does not make mistakes, but that gender is morphologically realized in more ways than our current binary dimorphic cultural frame allows us to recognize. The "fuzzy" model, then, replaces the overly simplistic emphasis on the poles at the extremes of a continuum, and offers a way for us to recognize the polar binary model as a linearized circle (a three-dimensional circle flattened out as a linear two-dimensional model) with a model that emphasizes and valorizes the "radical center." This is the 50% point: the point that problematizes and undermines any binary structure: the point at which A does = not A, at which either/or evaporates in the face of between-ness or doubleness.

The model I am proposing might look a little like a retreat to pre-Enlightenment categories. As Randolph Trumbach has demonstrated in his careful research into the gender and sexual identities of the eighteenth century, the Hermaphrodite was a recognized embodiment prior to the eighteenth century, but one that was erased from the Symbolic as a corporeal possibility by the middle of that century. Lorraine Daston and Katherine Park have noted a divergence in the late sixteenth century between two "distinct and, in many ways, contradictory theoretical traditions, Hippocratic and Aristotelian" (Daston and Park in Fradenburg and Freccero, 1996:117–136, esp. 118). While the Hippocratic model—associated with Galen—viewed hermaphrodites as between genders, embodying a point on a "sexual spectrum, ranging from unambiguously male...to wholly female;" the second model was rooted in Aristotle's *Generation of Animals*, where hermaphrodism was explained as excess matter from the mother, which exceeded that needed for one fetus but was not enough for two, producing an extra set of genitals. While the Galenic or Hippocratic model allowed hermaphroditism a place on the scale of natural being, the Aristotelian model designated hermaphroditism as excess, monstrosity, or accident, which needs surgically adapting to the dimorphic gender model.

The retreat from the binary that has dominated cultural forms in the modern period is an ethical project. It is not possible to ignore the binary, but it is possible to engender models for sexual embodiment that re-constitute normativity. I suggest that this process might not take as long as we fear it will, given the remarkable speed at which information technology is becoming the mechanism by which the Symbolic is maintained and reproduced. It strikes me that we might be at a moment that asks us to choose between these epistemic models, and it is for this reason that I foreground an ethical position in these arguments, and recommend a healthy suspicion of empiricism. *Means justify ends.* My political agenda is to produce a cultural model that includes and represents *all possible embodiments* of the human subject in a normative model, and then see where it takes us. I have tried to demonstrate in this chapter that in order to achieve that ethical/political end, I have to include intersex not as an added extra, but as the radical center to which all embodied subjects aspire.

References

Butler, Judith. (1990). *Gender trouble: Feminism and the subversion of identity.* New York: Routledge.

Butler, Judith. (1993). *Bodies that matter: On the discursive limits of "sex."* New York: Routledge.

Cixous, Hélène. (1975). Sorties. In Hélène Cixous and Catherine B. Clément, *La Jeune née.* Série *Féminin futur.* Paris. 114–246.

Cixous, Hélène. (1986). Sorties. In *The Newly Born Woman* (Theory and History of Literature, Vol. 24), ed. Hélène Cixous, Catherine Clement, and Sandra M. Gilbert. Minneapolis, MN: University of Minnesota Press.

Daston, Lorraine, and Katherine Park. (1996). The hermaphrodite and the orders of nature: Sexual ambiguity in early modern France. In Louise Fradenburg and Carla Freccero (eds.), *Premodern sexualities.* New York and London: Routledge.

Dupré, John. (1993). *The disorder of things: Metaphysical foundations of the disunity of science.* Cambridge, MA: Harvard University Press.

Epstein, J., and K. Straub (eds.). (1991). *Body guards: The cultural politics of gender ambiguity.* New York and London: Routledge

Fradenburg, Louise, and Carla Freccero (eds.). (1996). *Premodern sexualities.* New York and London: Routledge.

Kosko, Bart. (1993). *Fuzzy thinking: The new science of fuzzy logic.* London: Harper Collins.

Laclau, Ernesto. (1993). Politics and limits of modernity. In *Postmodernism: A reader*, ed. Thomas Docherty, New York: Harvester Wheatsheaf, 329–343.

McNay, Louis. (1994). *Foucault and feminism.* Cambridge: Polity Press.

Roberts, Gareth. (1994). *The mirror of alchemy: Alchemical ideas and images in manuscripts and books from antiquity to the seventeenth century.* London: The British Library.
Shapiro, Judith. (1991). Transsexualism: Reflections on the persistence of gender and the mutability of sex. In J. Epstein and K. Straub (eds.), *Body guards: The cultural politics of gender ambiguity.* New York and London: Routledge.
Trumbach, Randolph. (1991). London's Sapphists: From three sexes to four genders in the making of modern culture. In J. Epstein and K. Straub (eds.), *Body guards: The cultural politics of gender ambiguity.* New York and London: Routledge, 112–141.
Zadeh, Lofti. (1988). Fuzzy logic. *Computer,* April 1988, 83–93.
Zadeh, Lofti, and Janusz Kacprzyk (eds.). (1992). *Fuzzy logic for the management of uncertainty.* New York: Wiley.

Notes

1. Ernesto Laclau identifies Saussure's contribution to structuralism in his transformation of the linguistic model: "Saussure...tried to locate the specific object of linguistics in what he called langue, an abstraction from the ensemble of language phenomena based on a set of oppositions and definitions, the most important of which are: langue/parole, signifier/signified, syntagm/paradigm. The two basic principles that oversaw the constitution of the linguistic object were the propositions that there are no positive terms in language, only differences, and that language is form, not substance." Laclau, Politics and the Limits of Modernity, in Thomas Docherty (ed.), *Postmodernism: A Reader,* 329–343, esp. 332–333.
2. I am interested in the embodied hermaphrodite as a social subject, subject of the discourse of binary gender, and subject to the cultural injunction to approximate its corporeality to one of the gender prototypes. It is also worth considering the mythopoetic hermaphrodite, whose role it has been to suggest the beginning or end of gender. See especially Plato, *Symposium,* and Ovid, *Metamorphosis.* The alchemical hermaphrodite represented the last stage in the transformation, adjacent to the Philosopher's Stone itself. See Gareth Roberts (1994:84, 89) "The union of the principles had its social aspect in alchemical images of marriage... The consequences of the union might be dangerous...A more harmonious and stable state consequent upon union was often figured as a hermaphrodite, the alchemical 'Rebis.' It is possible to read the Judeo-Christian story of the Fall from Eden as the mythopoetic recording of the fall from a hermaphroditic unity into the polarized binary of sexual difference. In this reading, the Fall into consciousness is the Fall into consciousness of sexual difference (hence the fig leaves).
3. For a useful argument within the biological sciences against the essentializing of gender, see John Dupré (1993:68–80).
4. Lotfi Zadeh is credited with having "invented" fuzzy logic, and more specialized or courageous readers might like to try Zadeh (1988) and Zadeh and Kacprzyk (1992).

Fractured Masks:
Voices from the Shards of Language

Lee Anderson Brown

Introduction
Intersexuality—the phenomenon of people being born with ambiguous biological gender presentation[1]—is something that has been recorded throughout history. Further, it is something of a cliché to suggest that during that time it has been given a wide variety of meanings. In this paper I want to explore the modern experiences of intersexuals, especially the ways in which the experiences of intersexuals have been erased in the modern conceptual frameworks that are used to position them in this society, today; in other words, how the voices of intersexuals have been silenced and continue to be silenced by and in theoretical discourses of gender and sexuality. This endeavor is made more bitter by the knowledge that in these discourses intersexual people are "present" in their absence.[2] That is, they are often talked about but in such a way that intersexual people are not able to read themselves or their stories in these discourses. I will look at how this situation has come about, describe its current effects, and point to some future directions that may help to remedy it.

One question that has preoccupied me, and that was the inspiration for this chapter, is why there are so many transsexual and so few intersex activists, especially since even by the most generous figures, trans incidence in the population—given at 1 in 9,000 for Australia (Walters, 1997)—is significantly less than intersex (IS) incidence (given at 1 in 2,000 by IS activist groups), indeed less than a quarter of it. Such a discrepancy may be the result of some or all of the following reasons:

 * 1 in 2,000 includes many people whose genital ambiguity is slight or includes Androgen Insensitivity Syndrome (AIS) and Turner Syndrome people who don't necessarily feel outraged by their experience of the medical profession;
 * a number of people do feel "cured" by early medical intervention (although I've yet to meet one);

* many people have passed away due to complications from their condition, suicide, or some other accident/misadventure.

However, I find none of these "explanations" convincing and feel we need to look elsewhere to find the reasons for the absence—with a few notable exceptions—of intersexual activists or an activist network. In doing this it may be best to start at a significant moment in the development of modern medical practice, specifically, that period when forensic medicine slowly became the discipline of sexology as we know it today.

Hermaphroditic Substitutions

Proto-sexology, like medicine generally, seemed fascinated by intersexual people (the medical journals of the time record medical society meetings where intersexual corpses were dissected, and cases were discussed and argued over as to origins and diagnoses).[3] Of course there was nothing new here—intersex people had been studied, commented on, and assigned to their "correct" genders for many centuries previously. Indeed, by the nineteenth century the desire to attribute the true sex of a person was tuned to a fine art and seen as the duty of the medical and legal professions (see Laqueur, 1990:136). The example of Alexina Herculine Barbin is the obvious reference point. At one point in hir commentary on Alexina's[4] memoirs, Auguste Tardieu says,

> To be sure, the appearances that are typical of the feminine sex were carried very far in his case, but both science and the law were nevertheless obliged to recognise the error and to recognise the true sex of this young man. (quoted in the "Reports" section of Foucault, 1980:123)

At another point the doctor who first diagnosed Alexina's condition, Dr. Chesnet, after finding what zie considered enough markers of masculinity, declares "We can now conclude and say: Alexina is a man, hermaphroditic, no doubt, but with an obvious predominance of masculine sexual characteristics" ("Chesnet report" quoted in Foucault, 1980:128).

Significantly these quotes show that there is no "hermaphrodite" (in the social sense) in modern medicine,[5] only "hermaphroditic" —what was once a noun is now only an adjective—and the specificity of being born with ambiguous genitalia is subsumed unproblematically into the category of abnormal male or abnormal female. There is no doubt in the minds of the doctors that the person is on a particular side of the gender divide; the trick is to find it.

However, while hermaphroditic bodies had to be fixed there was a push to have hermaphroditic behavior recognized as a form of specificity based on a hermaphroditic essentialism (that is, neither/both sex-genders), especially through the work of Karl Heinrich Ulrichs. In hir work as an activist trying to achieve legal recognition for same-sex desire, Ulrichs coined the term "a female soul enclosed in a male body"[6] as the definition of a Uranian—hir term for homosexual. In other words, Uranism was conceptualized as a type of psychological hermaphroditic condition. Ulrichs' focus was on a psychic gender inversion or a hermaphroditism of the mind.

And if intersexual people were denied a right to a specificity, then that right—to be recognized as a third sex—was taken up and demanded by the new wave of homosexual activists of the period. This can be seen most clearly in the way the term "intersexual" itself was used. Coined, according to the *Oxford English Dictionary*, by J. Grote in 1876 to describe, in what seems to be a metaphysical way—that which the sexes have in common or share,[7] it did not take up its modern endocrinological meaning until R. Goldschmidt used it in a journal publication of 1915 and in the 1917 edition of *Endocrinology I*. In the *Endocrinology* entry zie says, "We have proposed the use of the terms intersexe, [sic] intersexual, intersexuality instead of sex-intergrades."[8]

Several years earlier "intersexual" was taken up as a synonym for psychic hermaphrodite by Xavier Mayne (a pseudonym) in hir 1912 book, *The Intersexes*, where zie argues that homosexuals are a third sex positioned psychically between male and female. This formulation of homosexuality—following Hirschfeld and Ulrichs—as gender inversion was very influential[9] and was a common line of argument of the time (Earl Lind also uses the same rationale [see Lind, 1975a, 1975b; Chauncey, 1994]).

However, while the term "intersexual" was soon superseded by "homosexual"[10] to describe sexual desire and acts between adults of the same gender, its use to describe people born with ambiguous genitalia grew and it became the dominant term for this phenomenon. It was still a pathological label in that one could only be intersexual as a condition one had, but one could not be an intersexual as a personal identity (much less a hermaphrodite except in "Side-Show Alley"[11] and its successor in the public imagination—the tabloid press.) One was either male or female with an underlying condition that made the determination of correct gender "difficult" but not "impossible."

Jorgenson—Thoroughly Modern Hermaphrodite

Now we quickly move to the 1950s and Christine Jorgenson. When news of hir sex reassignment became public there was some speculation in the media that it was a correction of a "biological" mistake

—in other words that zie was intersexual. When it was revealed that zie had presented with normal male genitalia then the media tide turned against hir in a big way (*Time*, 1953:41; *Time*, 1952:43, 49). In the public imagination Jorgenson moved into that abject space previously occupied by hermaphrodites—the freak-show spectacle of being neither man nor woman but both/neither.

Jorgenson's struggle to move away from this social recognition as transgressive hermaphrodite is important because it defined transsexuality for the next 30 years. Also, it was around this time that early childhood surgical intervention really started to be used to render intersexuals invisible,[12] and Jorgenson's story pivots around the way that the concept of hermaphrodite was transferred onto transfolk in the public imagination. If society had rid itself of "old-style" hermaphrodites by consigning them to the medical cabinets, then medicine was able to provide new hermaphrodites for public consumption.[13] The publicity surrounding Jorgenson's case borrowed and rewrote the hermaphroditic myth as read through the contemporary medical theories about the nature of gender difference. Indeed Jorgenson set the seeds for the "classic transsexual story" by writing hir story as a rebuttal of this perspective. With the backing of some aspects of the medical profession and because zie was so presentable—a charming, a normal, clean-cut, clean-living gal—zie tried to present transsexuality as the situation of the "girl next door" who was tragically and unfortunately trapped in the wrong body (Califia, 1997), in other words, rewriting and echoing the old formula used to discuss homosexuality.[14]

Writing the Self

Stories are important because it is by the stories we tell, and the way they are received, that we are able to "write" ourselves into existence. We all take or adapt the available language and, according to how the conceptual meanings of a language or word-set resonate for us, position ourselves within its framework in order to coexist in a given society. This process is dependent on the language being able to provide or be easily molded into a conceptual framework that is adequate to describe and express our life situation and feelings. As Seyla Benhabib says:

> The Self is both the teller of tales and that about whom tales are told. The individual with a coherent self identity is the one who succeeds in integrating these tales and perspectives into a meaningful life history. When the story of a life can be told only from the perspective of others,

then the self is a victim and sufferer who has lost control over her existence. (quoted in Plummer, 1995:145)

In Jorgenson's case, as well as those trans story-tellers who have come after hir, this story has flexed and changed to reflect the changing social circumstances of transfolk (Califia, 1997; Bates, 1997). Still, what was set down as the classic or canonical transsexual identity provided a framework in which one could begin to enter discourses about trans subjectivity. Even now, given all the current debate over the meanings of a trans identity—that is, transitional, third gendered, non-gendered or floating gendered—the classic story written by Jorgenson and others of that era provides an important starting point of self-definition, even as that point is ultimately opposed and rejected[15] by writers such as Kate Bornstein (1994) or Riki Anne Wilchins (1997).

Re-Writing Intersexual Stories
When we look at the possibilities of writing a "core" or "classic" intersex story we are confronted with a number of serious difficulties in trying to locate a central story by which we can understand our situation.

* Firstly, many intersexual conditions do not necessarily entail genital or gender ambiguity (mild forms of hypospadias associated with a range of conditions or with congenital adrenal hyperplasia [CAH]).

* Secondly, the diversity of genital ambiguity that is possible means that trying to locate a common experience of the body or feelings about the body are made very difficult.

* Thirdly, the story given to us by the medical profession—when indeed we are given one—is shrouded in concepts of lack, abnormality, incompleteness, and freakishness; if we are to believe the stories told about us, then we have to assume the mantle of the abject.

* Finally, if we try to reappropriate the social stories of hermaphrodites and androgynes in this society, we find those stories alien and strange. Because of the aspects I've discussed in this paper, even this story is no longer our own; it belongs to others and now has different reference points (for example, the mind rather than the body, or transgressive sexual nature).

However, there are some gaps and silences where some intersexuals are able to find common cause and begin the process of writing the self. The most important example has been finding parallels with, and adapting, the "genital mutilation" story. What began as a Western commentary and critique of non-Western religious cultural practice found resonance with the experience of intersexuals, and we have used it as a means to understand our feelings about the childhood medical treatment and surgery that many of us have undergone. In other words, the story of genital mutilation finally gives us one key into the language that can begin to explore our experience.

I believe another important entry point is to be found in exploring the psychologically traumatic nature of the silences, lies, and betrayals to which intersexuals have been subjected by this process. Because we were subjected to these medicalized procedures and therapies, without our consent or even knowledge, at an age when we were too young to do anything but "trust" those "looking after" us, and at the time it was "impossible" to hate them (not being able to understand the reasons for our treatment), we directed that discomfort and hatred inward toward ourselves. This profound level of trauma and violation during the formative years of childhood not only destabilizes our sense of ourselves but also makes it more difficult to recover from those traumas because we've rarely had an unproblematic sense of ourselves—deemed essential by Benhabib (1992:168)—as a reference or even as a starting point.

This leaves the last, probably the most important point. Our stories must focus on the healing processes that we are able to enact in our own lives or are able to borrow from other people's lives and experiences. The severe nature of the traumas we've faced means that the healing processes we are able to implement in our own lives must always have priority over anything else. Unless we are able to create a "safe space" for ourselves as a foundation from which we can venture out into the "battle-fields" of activism then we risk doing more damage to ourselves and not helping the "cause," those we love, those who love us, and ourselves (our true "cause").

It is these aspects—more than any notion of third or fluid gender identity—that I believe are crucial to the project of writing an "intersexual self" at this point of time. Maybe at some stage it will be possible for a number of intersexuals to write of the joys of being "both" sexed/gendered but now it is so important to rewrite those stories that have been written into our bodies, not at birth, but during our infancy and childhood by others. Until those stories are addressed it is difficult, if not impossible, to be what seems to be expected of intersexuals in this society (not the least by some gender theorists and some transfolk), that is, to be happy hermaphrodites. Indeed the way some transfolk read the hermaphroditic story back to intersexuals, ignoring the trauma we've been through, only adds to the trauma and drives intersexual people away.

Conclusion

Stories about the self are the main way that humans position themselves with regard to others and the way that they rationalize their behaviors, actions, and reactions. In this chapter I have looked at how one story based on the presence of intersexual people—hermaphroditism—has been taken by others and applied to a range of situations. While some people find comfort in these stories, when they are read back into the lives of intersexuals they seem foreign, strange, and for a large number, only tend to confirm our alienness. In the face of this I believe intersexual people need to focus on and privilege other aspects of their stories. Anyone who wishes to write an intersex story would be well advised to use the starting points suggested here rather than any utopian notion of gender fluidity.

References

Australian Medical Journal. (1914a). July 18, 3, 70.
Australian Medical Journal. (1914b). Nov 21, 21, 505.
Bates, Denise. (1997). Someone else's gender? Searching for the woman in transsexual autobiographies. Unpublished paper.
Benhabib, Selya. (1992). *Situating the self: Gender, community and postmodernism in contemporary ethics.* New York: Routledge.
Bornstein, Kate. (1994). *Gender outlaw: On men, women and the rest of us.* New York: Routledge.
Bullough, Vern, and Bonnie Bullough. (1993). *Cross dressing, sex, and gender.* Philadelphia: University of Pennsylvania Press.
Califia, Pat. (1997). *Sex changes: The politics of transgenderism.* San Francisco, CA: Cleis Press.
Chauncey, George. (1994). *Gay New York: Gender, urban culture and the making of the gay male world, 1890-1940.* New York: Basic Books.
Colapinto, John. (2000). *As nature made him: The boy who was raised as a girl.* London: Harper Collins.
Foucault, Michel. (1980). The dossier in *Herculine Barbin: Being the recently discovered memoirs of a nineteenth century French hermaphrodite.* Translated by R. McDougall. New York: Pantheon Books.
Hekma, Gert. (1996). "A female soul in a male body": Sexual inversion as gender inversion in nineteenth-century sexology. In *Third sex, third gender: Beyond sexual dimorphism in culture and history*, ed. G. Herdt. New York: Zone Books, 213–239.
The Lancet. (1872). June 15,.828–829.
Laqueur, Thomas. (1990). *Making sex: Body and gender from the Greeks to Freud.* Cambridge, MA: Harvard University Press.
Lind, Earl. (1975a [1918]). *Autobiography of an androgyne.* New York: Arno Press.
Lind, Earl. (1975b [1922]). *The female impersonators.* New York: Arno Press.

Mayne, Xavier [Pseudonym of Edward Prime-Stevenson]. (1908, 1975). *The intersexes: A history of similsexualism as a problem in social life*. New York: Arno Press.
Plummer, Ken. (1995). *Telling sexual stories: Power, change and social worlds*. London: Routledge.
Time. (1952). December 15, 43, 49.
Time. (1953). The case of Christine, April 20, 41–42.
Walters, William. (1997). The transgender phenomenon: An overview from an Australian perspective. *Venerology*, 10(3):147–149.
Wilchins, Riki Anne. (1997). *Read my lips*. New York: Firebrand Books.

Notes

1. Many intersexual conditions do not manifest themselves through ambiguous external genitalia. However, at least for the beginning of this paper, I chose to focus on those people born with ambiguous genitalia, partly because this is becoming the word's social definition.
2. See the "John/Joan" case (Colapinto, 2000) where a young male child was re-gendered as a small girl after accidental damage to his penis.
3. Both *The Lancet* (June 15, 1872:828–829) and the *Australian Medical Journal* (1914a:70; 1914b:505) published accounts of intersexual people in their first year of publication.
4. The name zie uses for hirself in hir memoir.
5. This is even true when the terms "true" or "pseudo-hermaphrodite" are used; what is "true" or "pseudo" is not the person but the person's gonadal tissue.
6. *Anima mulicbrus in corpore virili inclusa* (quoted in Hekma, 1996:219).
7. The *Oxford English Dictionary* quotes the following from Grote: "The intersexual feeling which belongs to all animals." It also includes a quote from Havelock Ellis' *Studies in the Psychology of Sex*, vol.1: "This is quite true of intersexual love."
8. It is interesting to speculate that zie chose the term to match the sexual endings of heterosexual and homosexual then gaining some currency, but without having read the original book it is impossible to know for sure.
9. Of course it was not the only one. Gert Hekma (1996:229) suggests that "Although the views of Hirschfeld and many other "heterosexual" physicians may have had the strongest social impact, attitudes within the homosexual movement were not clearly in favor of the third sex." The word "heterosexual" is in quotes because Hirschfeld is thought to have been homosexual although zie never publicly spoke of it.
10. After going through several other stages which I'm unable to discuss here.
11. Another place where the symbolism of the hermaphrodite has been and continues to be important in popular entertainment.
12. A *Time* article (1952:42) states that "sex transformation is far from a medical rarity, that similar cases occur in hospitals all over the U.S. right now," "sex transformation" meaning, of course, "genital correction" operations on intersexuals.

13 It's as though society needs to have a group in the position of hermaphrodite in order to help define what is normal by measuring itself against that deemed abnormal or outside the boundaries.
14 Another important point was that Jorgenson tried to put some distance between homosexuals and transsexuals although, as mentioned, using a theoretical perspective developed to describe homosexuality.
15 Similarly, cross-dressers have a specific contextual framework—heterosexual men who enjoy dressing in women's clothes—constructed by a small group of cross-dressers in an endeavor to encapsulate and frame their experience (Bullough and Bullough, 1993:280–286). It should be noted that this framework is also under challenge, especially by those who identify as gay or bisexual, or who wish to retain the right to opt for gender-changing medical procedures at a later stage in their life.

Conclusion

Tarquam McKenna

The manner in which homophobic and discriminatory practices are enacted because of heterosexist prejudice has implications for all humanity. The social, psychological, and political implications of this book, however, will remain academic as long as intolerance of difference continues to be practiced. In 1948 the newly formed United Nations passed its Universal Declaration of Human Rights. The declaration begins:

> The General Assembly proclaims this Universal Declaration of Human Rights as a common standard of achievement for all peoples and all nations.
> Article I
> All human beings are born free and equal in dignity and rights. They are endowed with reason and conscience and should act towards one another in a spirit of brotherhood.
> Article 2
> Everyone is entitled to all the rights and freedoms set forth in this Declaration, without distinction of any kind, such as race, colour, sex, language, religion, political or other opinion, national or social origin, property, birth or other status.

Despite changes in legislation aimed at protecting their rights, those who differ from the norms of gender are often discriminated against in practice. Harassment of gay, lesbian, transgender students typically goes from name-calling to unwanted sexual gestures to sexual touching—sometimes even rape. Many American students are refusing to tolerate homophobia in the form of battering and verbal abuse in schools. In November 1996, Jamie Nabozny won a $962,000 settlement from the Ashland, Wisconsin, public school system in a landmark decision, when a federal jury found that school officials had violated his constitutional right to equal protection. They had brushed aside his pleas for help during five years of escalating attacks that ended with his being kicked in the stomach so savagely that he required surgery. In the wake of the Nabozny decision, the U.S. Department of Education warned that schools risked losing their all-important federal funding if sexual harassment of gay students was permitted to flourish. In June, the department penalized Fayetteville, Arkansas, public schools for failing to protect Willie Wagner, a gay teenager who dropped out of high school after having his nose broken and his kidney bruised in the last of many attacks. Mark Iversen recently won a $40,000 lawsuit, and there have been other federal civil rights complaints aimed at the schools that have failed to protect their students (Price, 1999).

But there is also a less litigious appeal to ethical rights under a special International Bill of Gender Rights, modeled on the Declaration of Human Rights, first drafted in 1993, then reviewed, amended, and adopted in its current form at the International Conference on Transgender Law and Employment Policy (ICTLEP) in 1995. Though this bill is a theoretical construction with no force of law, it is recognized in principle by courts of law, administrative agencies, and international bodies such as the United Nations. Individuals are free to adopt the truths and principles expressed in it, and to lead their lives accordingly. In this fashion, the truths expressed in the bill will liberate and empower humankind in ways and to an extent beyond the reach of legislators, judges, and officials.

Long-term change in values comes from increased respect for all people, best initiated in a human context rather than a legal one. Most social institutions reinforce conventional "wisdoms" but schools in particular are responsible for educating the community for long-term changes in ethical behavior. The American Civil Liberties Union has set up a group in New York specializing in assisting gay youth. We still need to consider how we can best promote change to eliminate the misguided fear and loathing that transgender people speak of attracting in their daily lives. Discussion of the educational needs of androgynous people should be a priority as it is through education that negative attitudes to gender diversity can first be redressed. If a school principal were to introduce and enforce policies prohibiting discrimination against intersex, gay, lesbian, and transgender people, this would make schools safer. The gesture of one person in leadership stating that he or she will not allow discriminatory practices based on sexuality and gender is a powerful image.

The likelihood of such a stance against this type of gender discrimination in Western Australia, where this book originates, is minimal.[1] In this Australian state active discrimination against gay, lesbian, bisexual, intersex and transgender people is still endorsed. Because such people have been speaking out on the web, in social groups, antidiscrimination laws have been set in place in other

Conclusion 205

countries and are slowly beginning to have effect. However, many transgender people addressed in this book have remained invisible and the call of this chapter is twofold: to address invisibility and educate a heteronormative society to include transgender individuals and their social groups.

A primary task is to move people beyond their assumption of only two sexes and make them aware of the multiple possibilities of assigning sexes and genders. Androgynous people have been forced to move away from their true selves. The hidden selves need now to be moved to the place where the individual can be accommodated, tolerated, and affirmed. The multiple dimensions of the selves that we have observed in these chapters require acceptance of the unique nature of knowing that each person brings with their stories. The transgender, transsexual, and intersex people in this book are real living beings who experience joy, sadness, pain and delight. Being considered a "transgressive reality" excludes them from active and fulfilling life presences in society. Tolerance is the first step to understanding their worlds, but in keeping with a quality of life that is our right, these people deserve to be appreciated and affirmed.

Esther Pirelli/Esben Benestead (1998) writes of the quest of the transgender person as frequently being harsh and challenging. She states that it is not the child or adolescent's responsibility to take on the world but rather the world's role to take care of the transgender individual. Alain Berliner, the producer of the film *Ma Vie en Rose,* states that "the gender cards are being re-dealt." As borders coalesce and difference and diversity become acceptable the "pitfalls of sexual identity for the individual and for children in particular are still deep." Benestead calls for the goal of education (and medicine) to be the "rendering of the child/adolescent/young adult to a place of positive 'belonging.'" S/he asks the reader to consider positive belonging as being perceived and approved of by others in the same way as one perceives and approves of one's self.

This elusive, somewhat simplistic call to "belong" is the theme of the forgoing chapters. This book has asked the reader to call on places of positive belonging. In the past we have assumed that many aspects of existence for people who differ from the norms of gender were a "pathology." Transgender, transsexual, intersex, gay, and lesbian people do not suffer a pathology and because of this must not and cannot be, as Benestead states, "treated." Harry Benjamin in his recommendations for medical treatment of the intersex asks that terms such as "pathological" or "abnormal" be replaced by "unusual" or "different." Benestead is calling on the reader to consider the quality of engagement and encounter that is occasioned as life is lived for the transgender person. The narratives of the people throughout this text have been created as a reply to their "pain." As Martha Coventry relates her story of how when she asked her surgeon father direct questions about her sex, she was told she was "too much into self-examination," she says that the silencing caused as much pain as the surgery (quoted in Dreger, 1999:71–76). The pain of existence for androgynous people comes about because these people have been perceived as being "less than" who they really are.

It is time to take the challenge of enforced silencing and compulsory invisibility into classroom space. The manner in which we can do this is dependent not only on the policymakers and curriculum authorities but also on the exchanges that occur in the classroom. It is here that many malignant interactions occur and serve to bring pain to the students we teach. The young person who is perceived as "not belonging" is acted on by heteronormative rituals that are played out through power relationships which affirm heterosexuality as normal. The treatment that is needed is a relearning that must be undertaken by the class mates and teachers who need to know that there are other legitimate and meaningful ways of being. Intolerance when crossing the borders of meaning is the key to the pain into which many other gender expressions are locked.

Educational acceptance of diversity must now move to the realm of affirming those people who are transgender. The interrelated network of class teachers, community workers, principals, and policymakers can generate a more fulfilling learning place for the transgender person by networking with all colleagues at all levels of education. The goal is to remove the stigma or label that has created a sense of difference, which, in the main, is self-loathing. The internalized oppression of self comes because a person's gender expression and/or sexual orientation is not seen as "good." The cost of living with a stigma and the management of stigmatized identity is hyper vigilant and excessively diligent behaviors, for instance, Martha's two years of masturbation for fear of revealing her erotic orientation to women. The stress of hiding one's identity is extreme. Intimacy, which requires disclosure, is withheld. For the stigmatized minority groups addressed in this book, the restricted quality of relationships and behaviors demands sensitive and supportive redressing by all levels of the educational communities. The incidence of negative behaviors attributed to poor self-concept and self-loathing include substance abuse, suicide, depression, and other high-risk behaviors.

The chapters have examined how living with a different concept of gender or sexual orientation agitates and confronts the entire community. Anti-homosexual theories are still being applied to the understanding of gay identities and to "treatment" approaches for gay people. Social attitudes toward androgynous people will of necessity be influenced by a misleading conflation of intersex, transgender, gay, and lesbian people. The complexities of sex and gender attributes require educators to think about why they want to discriminate on the basis of sex and gender. Gender equity consists of treating equals equally and unequals unequally in relevant respects. Understanding the complexity of gender requires us to go beyond a simple differentiation of males and females and constantly consider what makes differences relevant to particular purposes.

A POLICY FOR POSITIVE CHANGE AND RECOGNITION OF INTERSEX, TRANSGENDER, GAY, AND LESBIAN PEOPLE IN EDUCATION

These policies and recommendations for action offer possibilities for an inclusive anti-homophobic and androgynous positive learning and educational environment.

Providing Appropriate Social and Emotional Care in Education

Sensitivity must be demonstrated in redressing misinformation on gender diversity and the pathologizing that the media insensitively foster. The school community needs to learn that the historical legacy of psychiatric "cures" for transgender, gay, intersex and lesbian students' sexuality is fallacious and contributes to an atmosphere of suspicion doubt, and distrust for transgender, intersex, gay, and lesbian students and staff.

Securing Equality of Opportunity

Androgynous members of a school community need to see explicit indications that the school values gender equality. Schools must assign and generate policies to redress discrimination on the basis of "sexual orientation" especially where this has been a tacit way of discriminating against androgynous students and teachers. Academic policy must also express a commitment to gender equity for all.

Creating a Sheltered and Safe Environment

If physical and verbal harassment of androgynous students and teachers is considered the norm, then the school community has a responsibility to take pro-active measures to address this violence. The school community must ensure that androgynous students and teachers can live out their various identities and social roles without any fear of harassment.

Providing Androgynous Role Models

Studies consistently show that personal acquaintance with gay and lesbian people is the most effective way of reducing homophobic bigotry. There are, however, no studies to illustrate the accommodation of androgynous people within school settings. Androgynous students would benefit from having explicit and positive role models to redress inaccurate stereotypes. Rituals of acceptance must be established, as schools create the conditions necessary for transgender, gay, and lesbian students and faculty to feel safe in "coming out" and being themselves.

Providing, Promoting and Advocating Support

Androgynous students and staff require affirmation and acceptance. This is a move beyond tolerance. The school community can provide and promote agencies that redress the feeling of not belonging. Social, sporting, and recreational agencies can create an atmosphere of appreciation regardless of sexual orientation. Support services available to all students must address the specialized needs of the androgynous students and staff with sensitivity and openness.

Training Non-Androgynous Faculty and Staff

School support staff must become skilled in addressing the needs of androgynous students and staff. The expectation of being able to understand the social, emotional, and psychological needs of transgender, gay, and lesbian people must sit comfortably alongside the acceptance of the diversity of learning styles afforded most people. The unspoken assumption that all androgynous people are the same must be removed so that the diversity of this population can be acknowledged. Over-generalized ways of interacting with and encountering intersex, transgender, gay, or lesbian people must be examined and eliminated.

Homophobia and noninclusive teaching practices are insidious in their origin and ongoing staff training and development must redress the imbalance that currently exists. The school community could demonstrate that this training is a necessity and not a mere "add-on" or token endeavor to address the oppression of androgynous people. It involves dissemination of stories such as those that appear in this book.

Reassessing the Curriculum
Academic staff, lecturers and teachers, must consolidate existing curriculum practices to include transgender, gay, lesbian, and intersex issues and redress the tacitly assumed heterosexism in the curriculum. Heteronormative principles embedded in the curriculum, particularly in science education, especially the presumptions of the binary male/female which pathologizes androgynous people, must at least be investigated if not removed. Australian researchers are starting to recognize the nature of "only heterosexual" curriculum biases (Letts, 1998). Curriculum initiatives must be developed to expose people to the issues of identity formation that are unique to transgender, gay, lesbian, and intersex students—thereby removing this group from being identified as marginal.

Falsely Assuming Only Heterosexuality
Invisibility is devastating for transgender, gay, lesbian, and intersex people. People are often surprised at the number of gay, lesbian and transgender people living invisibly and often in a ghetto situation, within their immediate vicinity.

Breaking the Silences
Everything changes when we tell the tales of our lives. The raw material of existence for androgynous people is the flux of experience that they have had. The telling and retelling of their tales is essential to the making of inner meaning. The actions that have made us more intelligent have been preceded by dialogue, conversations, or listening to each other. The struggles of being that are noted throughout this text arise as people have been re-voiced. "Tell me a story," says the teacher to the young children in her class. "Can you give me an example?" Androgynous people have stories to tell and tell again. The awakening of their own inner eyes to self-reflection requires the openings to dialogue to being heard. To be heard these people need to converse with an other. The other may be the teacher or their own community. Breaking the silences will lead to their healing.

Conclusion
The pragmatics of change, which can invoke positive opportunities for transgender, intersex, gay, and lesbian people, must include active teaching against homophobic stances and value systems. The overuse of the binary definitions of being (homo/hetero: male/female) serves to exclude and in educational research must be seen as invalid. There is no foundation for assuming the existence of only male or female

and the adoption of only these two polarities is tantamount to neglect. The validity of research that excludes people who are both or neither male nor female is not a mere philosophical question. The question of research, which brackets off this group of silenced participants in the educational community, is a moral one. Gender stereotyping has been described as the way to move into entrenched value stances in education. If instead teachers move the learning culture away from the stereotype, more awareness is inevitable. With awareness of the multiplicity of genders and the expressions of sexualities comes more "truth" and meaning for all of humanity. The right to be included is central to all people. The human rights afforded to all people apply equally to androgynous people, though it may need a Bill of Gender Rights to confer such rights.

The individual educator has a great opportunity to redress the gender expressions that are examined in this volume. The move away from heteronormative assumptions must occur, and thereby the educator can move into action. The sympathetic interpreter of the worlds of androgynous people needs to bring change, through reactions to the homophobic rituals and ignorant discourse still prevalent in most social institutions.

References

Dreger, Alice Domurat. (1999). *Intersex in the age of ethics*. Hagerstown, MD: University Publishing Group

General Assembly of the United Nations. (1948). *Universal Declaration of Human Rights*. <http://www.udhr50.org/UDHR/udhr.HTM>

Letts, Will. (1998). The heteronormative practices of science education: Addressing curricula and pedagogies in schools. Unpublished paper presented at the National Conference of the Australian Association for Research in Education, Adelaide (LET98109).

Pirelli, Esther/Esben Benestead. (1998). Networking for the future of possibly transgendered individuals. Workshop presented at the Third International Congress on Sex and Gender, Exeter College, Oxford University.

Price, Deb. (1999). Tide turns on abuse of gay students. *The Detroit News*, January 18.

Notes

1. Western Australia has thus far neglected to act justly and fairly under the terms of the International Labour Convention No. 111. The Human Rights and Equal Opportunities Act (1990) of the Australian Government specifically included sexual preference as an unlawful ground for discrimination. The experience of gay, lesbian, and transgender Western Australians is that others have made their sexuality an issue in areas such as employment, education, accommodation, and the provision of goods and services. Western Australia has the highest age of consent for gay men in the world today and the worst anti-gay laws in the nation. It is the only state where consenting gay sex between adults is a criminal offense and offenders can be jailed for up to five years. (http://www.democrats.org.au/democrats/campaigns/sexuality/hhbil.html January 22, 1999)

Appendix: Internet Links and Resources

AEGIS—American Educational Gender Information Services:
http://www.ren.org
AIS Support Group (Australia):
http://www.vicnet.net.au/~aissg/
Centre for Hormone Research in Australia:
http://www.rch.unimelb.edu.au/hormone/
The Gender Centre (Australia): http://www.gendercentre.org.au
Genetic Alliance: http://www.genetic alliance.org/job.html
GIRES—Gender Identity Research and Education Society:
http://www.gires.org.uk/
GLSEN—Gay, Lesbian and Straight Education Network:
http://www.glstn.org/
Hermaphrodite Education and Listening Post:
http://users.southeast.net/~help/sexdiff.html
IFKSSG—International Federation of Klinefelter's Syndrome Support Groups: http://members.aol.com/kscuk/IFKSSG.htm
IFAS—The International Foundation for Androgynous Studies:
http://www.ecel.uwa.edu.au/gse/staffweb/fhaynes/IFAS_Homepage.html
International Foundation for Gender Education: http://www.ifge.org/
International Journal of Transgenderism:
http://www.symposion.com/ijt/ijtintro.htm
The Intersex Society of Australia: isozvic@hotmail.com
The Intersex Society of North America (ISNA): http://isna.org
ISGI—Intersex Support Group International:
http://www.symposion.com/ijt/ijtintro.htm
It's Time Illinois: http://ItsTimeIL.org
Klinefelter's Syndrome Association (in UK):
http://members.aol.com/kscuk/IFKSSG.htm
Murdoch Children's Research Institute:
http://murdoch.rch.unimelb.edu.au/
New Zealand Sex Chromosome Society:
http://homepages.ihug.co.nz/~nzkline/xxyguide.htm
New Zealand Transsexual Men's and Women's Resources:
http://nz.com/NZ/Queer/Trans-NZ
NSGC—National Society for Genetic Counsellors:
http://www.nsgc.org/index.html

OIGS—Outreach Institute of Gender Studies:
http://www.cowart.com/outreach/
The P.E.R.S.O.N. Project:
http://www.youth.org/loco/PERSONProject/
Press for Change: http://www.pfc.org.uk.html
Pride, South Africa: http://www.sapride.org/
Pride London: http://www.pridelondon.org/index.html
Pridenet: http://www.pridenet.com
The Pride NY Gay Lesbian Web Directory:
http://prideny.simplenet.com/index.html
Renaissance Transgender Association (Inc): http://www.ren.org/
Transgender Australia: http://www.pridenet.com/austral.html
Transsexualism: The Current Medical Viewpoint 2nd ed 1996:
http://www.pfc.org.uk
Transsexual Menace: http://www.helenavelena.com/tsmenace/
Transsexual Women's Resources: http://www.annelawrence.com./
Ultrasex [Beyond Division]:
www.cat.org.au/ultra/ultra1.html
UKIA—The United Kingdom Intersex Association:
http://www.ukia.co.uk/
XXY: http://www.globalwebsol.com/vv/index.htm
Youth Pride Alliance: http://www.youthpridedc.org/index.html

Glossary

Adrenal Hyperplasia (also called Congenital Adrenal Hyperplasia, or CAH). The most prevalent cause of intersexuality amongst XX people, with a frequency of about 1 in 20,000 births. An inherited enzyme deficiency (usually 21-hydroxylase or 11-hydroxylase deficiency), results in malfunction of the fetus's adrenal gland, and causes the overproduction of androgens. A female can fully masculinize to the point where a person with XX chromosomes looks like a boy with undescended testes. Because the virilization originates metabolically, masculinizing effects continue after birth. As in progestin-induced virilization, sex phenotype varies, with the possible added complication of metabolic problems (salt loss). The metabolic effects of CAH can be counteracted with cortisone.

Ambiguous Genitalia. Any condition in which the genitalia do not conform with those of normal males and females, such as hypospadias, undescended testes, fused labia, clitoromegaly, micropenis, combination of penis and vagina (5%–15% of the population).

Androgen. A generic name for male sex hormones, including testosterone and androstenedione.

Androgen Insensitivity Syndrome (AIS). A genetic condition, occurring in approximately 1 in 20,000 individuals. With AIS, the body's cells fail to respond to testosterone and cannot absorb either fetal or pubertal androgens. Those with the complete form of AIS are born with XY (male) chromosomes, internal testes, and female genitals. Their vaginas may be short or virtually absent, which is related to their absent uterus and cervix. However, because cells fail to respond to testosterone, there are no epididymis, vas deferens, or seminal vesicles. Without medical intervention, these children usually experience feminine puberty, because testes produce estrogen as well as testosterone. For this reason, AIS used to be labeled "testicular feminization." AIS individuals are generally raised as girls. At puberty, the estrogen produced by the testes produces breast growth, though it may be late. The individual does not menstruate and is not fertile. Most AIS women have no pubic or underarm hair, but some have sparse hair. When an AIS child is diagnosed during infancy, physicians often perform surgery to remove her undescended testes, because of the risk of cancer.

Androgynes. Individuals whose assumed characteristics are not limited to either of the two traditionally accepted gender classifications, masculine and feminine. Androgyny can include a variety of experiences including androgynous presentation, physique, behavior, wardrobe, and social roles.

Androgynous. Usually applied to a person whose gender identity is not obviously male or female.

Androphilic Female. A female whose erotic orientation is to males.

Androphilic Male. A male whose erotic orientation is toward males.

Aposthia. The condition of being born without a foreskin.

Autogynephilia The performativities of those who are sexually aroused by the thought or image of themselves as the opposite sex.

Balkan Sworn Virgins. A group of women in Eastern Europe who take on the role of men with concomitant celibacy, when there is a shortage of men.

Being Read. A slang term for being identified by the birth gender.

Berdache. Members of a North American Indian tribe who are assigned a gender that is neither male nor female. The gender may accommodate homosexuals or intersex persons but seems mostly to ascribe specific gender roles.

Bigendered. One who has a significant gender identity that encompasses both genders, masculine and feminine. Transsexuals normally do not consider themselves to be bigendered.

Biological Sex. Being male or female, as determined by chromosomes, or external genitalia, or sexual reproductive organs, depending on historical context.

Bisexual. One whose sexual orientation is toward both males and females.

Brain Sex. The physical "sex" of the brain. Based on the theory/evidence that the human brain is different for genetic males and genetic females.

Butch. Masculine or macho dress and behavior, regardless of sex or gender identity.

Camp. To exaggerate feminine behaviors, usually for others' entertainment. Also, "to camp it up."

Clitoromegaly. Having a clitoris longer than the acceptable female medical standard length of 0.9 cm. Infant genital appendages between 0.9 cm and 2.5 cm in length are unacceptable by medical standards.

Clitoridectomy (or **clitorectomy**). A surgical technique involving removal of the clitoris.

Clitoroplasty. Any surgery on the clitoris, usually to reduce its size.

Compulsory Heterosexuality. The assumption that a standard of heterosexuality exists, and that all divergence from that standard is impaired or deviant behavior.

Cross-Dressing. The adoption, fully or partially, of the clothes normally identified as belonging to the opposite sex. People may cross-dress for a variety of reasons of which transvestism, transsexualism, and fetishism are the commonest. Some people may also cross-dress as part of a disguise or for entertainment. Others may cross-dress as part of masochistic activities.

Cross-Dresser. One who, regardless of the motivation, wears the clothes, makeup, etc., assigned by society to the opposite sex. Generally, these cross-dressers do not alter their bodies. Though the clinical term for such people is "transvestite," most of them prefer the term "cross-dresser."

Cross-Living. Living full-time in the preferred gender image, opposite to one's assigned sex at birth, generally in preparation for a sex-change operation.

Cryptorchidism. Literally, a hidden testicle. It includes true undescended, ectopic and absent testicles.

DES. *diethyl stilbestrol.* A synthetic estrogen often prescribed to prevent miscarriage and morning sickness resulting in an abnormally high frequency of intersex characteristics in the genitalia of DES-exposed boys, and a higher than average incidence of depression and other psychoses in DES-exposed adults.

Drag Queen. Generally a MtF cross-dresser who usually goes to wild extremes, whether toward a glamorous or campy end, often for other people's entertainment, for appreciation or for shock value. This term is considered derogatory by some.

Diagnostic and Statistical Manual of Mental Disorders **(DSM IV).** The guidelines published by the American Psychiatric Association detailing what is and is not a psychiatric illness. Transsexualism and transvestism are included in the list of psychiatric disorders, as is gender identity disorder. Homosexuality was removed in 1979.

Electrolysis. Process of killing hair follicles, especially of facial and neck hair, usually with an electric needle.

En Femme. Projecting one's person to society as a female through clothing and mannerisms (i.e., dressed as a woman).

En Homme. Projecting one's person to society as a male through clothing and mannerisms (i.e., dressed as a man).

Epicene. While this word has taken on a variety of figurative meanings over the centuries (Ben Jonson used it to mean something like "effeminate"), it emphasizes what is common to both sexes. Its Greek root means "common," and it shows up in descriptions of garments either sex can wear, or places both sexes dwell. It is probably preferable to "androgyne."

Erotic Orientation. A medical term for sexual preference.

Estrogen. Generic name for one of the main groups of so-called female sex hormones (UK. spelling "Oestrogen"), though it is present in all sexes.

FtM or F2M. = Female-to-Male. Used to specify the direction of a change of sex or gender role from female to male.

Fa-fa'fine. A Samoan practice in which males can live the social role of women freely and without any shame. They often carry out traditional feminine roles such as child care, cooking, and cleaning without the need to cross-dress, though some take up jobs as female entertainers and singers.

Female Impersonator. A male who on specific occasions cross-dresses and employs stereotypical feminine dialogue, voice and mannerisms for the entertainment of other people (see **Camp**).

Female. One of the two physical sexes. Normally based on the primary sex characteristic of having a vagina. (see **Primary Sex Characteristics**). Legally, anyone with a female birth certificate.

Female Pseudo-Hermaphroditism. The condition in an XX person of having ovaries and "male" genitals. It represents about one-third of all cases of intersexuality.

Feminine. The gender role assigned to females.

Femme. Feminine or effeminate dress and behavior, regardless of sex or gender identity (see Butch).

Ferm. A name suggested by Anne Fausto-Sterling for a female hermaphrodite, that is, one with XY chromosomes but with a female phenotype including a vagina, sometimes a uterus and an enlarged clitoris.

Fetishistic Transvestite. A transvestite whose primary cross-dressing motivation is erotic response.

Gay. The most common term used for homosexual men, originally used derogatorily, but now an identifying term.

Gender. Gender is expressed in terms of masculinity and femininity. It is how people perceive themselves and how they expect others to behave. It is largely culturally determined, with the apparent exception of gender dysphoric persons.

Gender Dysphoria or **Gender Identity Disorder.** A psychiatric condition first identified at Johns Hopkins University in 1978. Refers to the dissatisfaction with one's gender (masculinity or femininity), which is in conflict with one's physical sex. In many countries, one must provide evidence of gender dysphoria before sex reassignment surgery is possible.

Gender Community. Colloquial for transgender community. People who identify as not having a gender identity that matches society's rules for their birth-physical sex, or those who identity with the gender community.

Genderfuck. Deliberately sending mixed messages about one's sex, usually through dress (e.g., wearing a skirt and a beard). Deliberate flaunting of gender norms with a goal of shocking others.

Gender Identity. The gender to which one feels one belongs.

Gender Neutral. Clothing, behaviors, thoughts, feelings, relationships etc. which are considered appropriate to both genders/sexes.

Gender Outlaw. (Coined by Kate Bornstein.) A person who defies traditional gender roles.

Gender Reassignment Surgery (GRS). Term used in the U.K. for Sex Reassignment Surgery (SRS).

Gender Role. Cultural expectations of behavior as appropriate for members of each sex, relative to location, class, occasion, history, and numerous other factors.

Genetic Girl. Female at birth regardless of one's present sex or gender identity (also known as Genetic Woman or Genetic Female).

Genetic Male. Male at birth regardless of one's present sex or gender identity.

Genetic Sex. The presumption that the XX (female) or XY (male) chromosome pair most heavily influence primary sex characteristics and therefore define one's "genetic" sex.

Genitoplasty. Any surgery on the genitals (female or male).

Gonad. The gland (ovary or testes) that produces eggs or sperm and hormones.

Gonadal Dysgenesis. A form of intersexuality characterized by undifferentiated gonads, sometimes resulting in atypical external genitals. It represents about one-third of all cases of intersexuality. Mixed gonadal dysgenesis (MGD) is characterized by an undeveloped gonad on one side and an abnormal testis on the other.

Gynocophilic Female. A female whose erotic orientation is toward females.

Gynocophilic Male. A male whose erotic orientation is toward females.

Haciendo caras. Putting on a face or mask—a performativity.

Herm. A name recommended by Anne Fausto-Sterling for a male hermaphrodite, that

Glossary

is, one with XX chromosomes but with a male phenotype including micropenis.

Hermaphroditism and **Intersexuality.** Where the physiological sex is ambiguous. A true hermaphrodite or a true intersex person is medically defined as someone with both ova and testes, though the combination may take different forms. Pseudo- or false hermaphrodites exist where other physiological combinations of male and female are present at birth, due to chromosomal complexes, such as Turner or Klinefelter Syndromes, or congenital errors of metabolism such as Androgen Insensitivity Syndrome and Adrenogenital Syndrome. It may also be caused by the hormone balance in the fetus or the placenta.

Heteronormativity. The process of assuming that everything and everyone conforms to a heterosexual norm, which influences education, public policy and interpersonal relations. Reference: Australian National Union of Students "Sexuality/Queer Policy" p. 149

Hjiras. A special caste in India for those who are infertile and neither male nor female, including eunuchs and intersex people.

Homophobia. Hostility to or distrust of anybody who is queer.

Homovestite. A person who obsessively, compulsively, and neurotically wears the clothing of their own sex.

Hormones. Hormone therapy is used by transsexuals to change some secondary sex characteristics, including breast size, weight distribution, and hair growth. Like most aspects of human chemistry, the endocrine system—which controls the body's production and balance of hormones, including sex hormones—is still, at best, only loosely understood.

Hormonal Reassignment Therapy. The introduction of the body to the hormones that affect the secondary sex characteristics of a transsexual. Estrogen is taken by MtFs and testosterone by FtMs. Intersex hormonal treatment can vary, but rarely are intersex people given a choice.

Hypospadias. A condition in males where the urethral meatus ("pee-hole") is not located at the tip of the penis, sometimes giving the appearance of ambiguous genitalia. Infertility may be present in the more extreme forms of hypospadias, where the testes are irregular and cannot produce viable sperm, but is not directly linked to the hypospadias. Surgery is undertaken to counter the risk of urinary tract infections or for cosmetic reasons, to overcome the trauma of having a penis that looks different. Suzanne Kessler *Lessons from the Intersexed* (1998) claims that the physical damage and emotional trauma of genital surgery are frequently far worse than the hypospadias itself.

Identity. How one thinks of oneself, as opposed to what others observe or think about one.

Intersex. The condition of having genital, gonadal, or chromosomal characteristics that are neither all "female" nor all "male."

Internalized Homophobia. The unconscious fear and hatred of homosexuality as experienced by lesbians and gay men themselves.

Kallmann's Syndrome. A form of hypogonadotrophic hypogonadism which can affect both men's and women's ability to go through puberty and can usually be treated by hormone replacement therapy. Kallman's is peculiar because it is associated with an absent or abnormally low sense of smell.

Kathoey. A Thai term meaning neither male nor female and applied to transsexuals, transvestites and homosexuals.

Kinsey Scale. A scale rating sexuality, from 0 being pure heterosexual and 6 pure homosexual. The Kinsey study results showed a bell curve with most people somewhere in the middle.

Klinefelter Syndrome. A quite common chromosomal variation, 47XXY, occuring in 1/500 to 1/3,000 births, considered by some to be an intersex condition. The only characteristic that seems certain to be present is small, very firm testes, and an absence of sperm in the ejaculate, causing infertility. Except for small testes, those with Klinefelter Syndrome are usually born with normal male genitals. But their testes often produce lower than average quantities of testosterone, so they don't virilize (develop facial and body hair, muscles, deep voice, larger penis and testes) as strongly as other boys at puberty. Many also experience some breast growth at puberty and some have a uterus and ovaries as well.

Label. How someone else thinks of one, as opposed to how one sees oneself (see **Identity**).

Lesbian. A woman whose sexual orientation is toward other women.

Mahu. A traditional Hawaiian term for MtF transgender individuals.

Mak nyah A Malaysian term for male-to-female transgendered people.

Male. One of the two physical sexes. Normally based on the primary sex characteristic of having a penis.

Male Impersonator. A female who, on specific occasions, cross-dresses and employs stereotypical masculine dialogue, voice, and mannerisms for the entertainment of other people.

Male Pseudo-Hermaphroditism. The condition in an XY person of having (usually undescended) testes and "female" genitals. It represents about one-third of all cases of intersexuality.

Man. One who identifies with the masculine gender role, regardless of present sex or sexual identity.

Masculine. The gender role assigned to males.

MtF or M2F. Male-to-Female. Used to specify the direction of a change in sex or gender role.

Metoidioplasty. A technique that transforms the transsexual's clitoris into a penis by first treating it with testosterone and then cutting the suspensory ligaments that hold the clitoris under the pubic bone.

Micropenis. A penis shorter than the male medical standard of 2.5 cm. Infant genital appendages between 0.9 cm and 2.5 cm are unacceptable to most doctors.

Muxe. A Mexican term for people who cross the male-female divide regardless of birth sex.

Neuter. One who has neither a penis nor a vagina (see Primary Sex Characteristics).

Noonan's Syndrome. The symptoms of Turner Syndrome with normal 46XX or 46XY karyotype. Also called Pseudo Turner, Male Turner Syndrome, or Ullrich syndrome.

No-op or Non-op. A person who has had all the hormonal/surgical treatments, except genital surgery, and who either has no desire to proceed with the surgery or cannot proceed due to financial constraints. GIDAANT is another term for this (Gender Identity Disorder, Adolescent or Adult onset, Non transsexual).

Passing. The opposite of "Being Read." A term often used to describe your "natural" ability to be accepted by most people as your preferred gender.

Per. An a-gendered term suggested by Christie Elan-Cane to replace gender-loaded pronouns.

Performativities. One's social presentation, including the way one walks and dresses and the roles one undertakes.

Physical Sex. The sex the body mainly matches, i.e., male, female, hermaphrodite, or neuter.

Post-op. Post-operative (after SRS) transsexual. May wish not to be considered a transsexual at this point.

Pre-op. Pre-operative (before SRS) transsexual. Normally implies the individual is planning SRS (see **No-op**).

Primary Sex Characteristics. Those primary physical characteristics that society uses to separate the sexes, usually the penis (male) or vagina (female).

Progesterone. One of the female sex hormones.

Progestin-Induced Virilization. A form of intersexuality caused by the mother's ingestion of synthetic androgens during pregnancy. Progestin is converted to an androgen (virilizing hormone) by the prenatal XX person's metabolism, causing virilization of the clitoris (ranging from enlargement of the clitoris to the development of a complete phallus and the fusing of the labia). In all cases ovaries and uterus or uterine tract are present, though in extreme cases of virilization there is no vagina or cervix, the uterine tract being connected to the upper portion of the urethra internally. The virilization only occurs prenatally and the endocrinological functionality is unchanged, i.e., feminizing puberty occurs due to normally functioning ovaries.

Queer Sexualities. Like "gay," a label assumed with pride by those whose sexual boundaries are perceived to transgress the normal. More often applied to sexualities such as gay, lesbian, or bisexual.

Read. When someone detects you are transgendered; also "clock(ed)."

Real Life Test. That period (usually a minimum of 1 year) imposed on the individual by the medical community in which he/she is required to live full-time in the role of the opposite sex before sex reassignment surgery.

Reproductive Organs. Testes or ovaries/womb: according to Alice Dreger *Hermaphrodites and the Medical Invention of Sex* (1988), the organs primarily used in the medical profession in the 1920s to differentiate men from women.

Sexual Orientation. Refers to the type of person to whom one is affectionally and sexually attracted, usually based on gender and sex characteristics. Also called erotic orientation. Orientation is *not* the same as behavior or preference.

Secondary Sex Characteristics. Facial and body hair, vocal timbre and range, breast size, weight distribution.

Sex Reassignment Surgery (SRS). A surgical procedure designed to change one's primary sexual characteristics (genitalia) from those of one sex to those of another (penis to vagina, or vagina to penis). May also include secondary

surgery such as breast implants or removing the Adam's apple.
She-Male. A popular—if often demeaning—term, generally used by non-transgender males seeking sex, to describe pre-SRS MtF transsexuals. MtF cross-dresser with "tits, big hair, lots of make-up and a dick" (Bornstein). Also, "chick with a dick," she-he or trangenderist.
Sissy Boy. A biological male who prefers to adopt a stereotypically feminine or neuter gender role, including dress but also domestic duties.
Standards of Care. The guidelines established in 1974 by the Harry Benjamin International Gender Dysphoria Association, as the minimum guidelines for a Transsexual Physical and Psychological Transition.
Testicular Feminization. The opposite of progestin-induced virilization.
Tranny. A popular term, not derogatory, used in Britain, Australia, and New Zealand to refer to a transgender individual.
Transgender. A term used in the U.S. to include transsexuals, transvestites, intersexuals, gender dysphorics, and cross-dressers. It can also represent a person who, like a transsexual, transitions—sometimes with the help of hormone therapy and/or cosmetic surgery—to live in the gender role of choice, but has not undergone, and generally does not intend to undergo, SRS (see **No-op**).
Transsexuals. Persons who feel a consistent and overwhelming desire to transition and fulfill their lives as members of the opposite gender. Most transsexuals actively desire and complete Sex Reassignment Surgery.
Transvestite. The clinical name for a cross-dresser. A person who dresses in the clothing of the opposite sex. Generally, transvestites do not alter their body.
Turner Syndrome. A child born with 45X0 karyotype, looking like a female, but completely lacking testosterone. In 90% of cases, there are no ovaries present causing absent menses and infertility in adult life, but 8%-16% of all cases are a mosaic of 45X0/46XX or 45X0/46XY, with less extreme symptoms.
***Winyanktecha* (*Wintke*).** Lacota Indian word meaning gender-crosser—literal translation, "two-souls person."
Woman. One who identifies with the feminine gender role, regardless of present sex or sexual identity.
5-Alpha-Reductase Deficiency. A form of androgen insensitivity caused by a genetic enzyme disorder that prevents testosterone from "masculinizing" the XY fetus's genitals before birth. The genitals "masculinize" at puberty.

Contributors

Lee Anderson Brown is classified as intersex and has undergone a series of surgeries to correct a hypospadias that many see as a physical deformity. He has completed a first class honors degree in sociology analyzing accounts of intersexuality at the University of Sydney and is now completing a doctorate on a theory of intersex, in the Department of Sociology and Social Anthropology.

Adrianne Dana-Tabet studied at Brandeis University in areas including ethnographic investigations of a variety of sexual and gender subcultures including sado-masochistic practitioners and a study of gay geography in the Boston area. Her current interest focuses on the critical analysis of the process of conducting fieldwork.

Richard Ekins is Director of the Trans-gender Archive and Reader in Social Psychology and Psychoanalysis at the University of Ulster, Coleraine, Ireland. He has published widely on various aspects of transgender, is the author of *Male Femaling: A Grounded Theory Approach to Crossdressing and Sex-Changing* (Routledge, 1997), and co-author of *Centres and Peripheries of Psychoanalysis* (Routledge, 1994) and *GenderBlending* (Routledge 1996).

Michael A. "Miqqi Alicia" Gilbert is Professor of Philosophy at York University, Toronto. S/he published the second edition of *How to Win an Argument* in 1997 (John Wiley and Sons,) and the novel, *Office Party* (Simon and Schuster, 1981). S/he also wrote the screenplay, *Hostile Takeover*, based on *Office Party*, which was produced by SC Entertainment in 1989. More recently s/he has been publishing scholarly articles in the areas of argumentation theory and gender theory in journals such as *Argumentation, Inquiry, Philosophy of the Social Sciences,* and *Informal Logic*. His/her book *Coalescent Argumentation* was published in 1997 by Lawrence Erlbaum Associates. Miqqi Alicia is a lifelong cross-dresser, and an activist in the international transgender community. S/he is editor of *The Monarch Reader*, published by the club Xpressions in Toronto, and a regular contributor to *Transgender Tapestry,* the magazine of the International Foundation for Gender Education.

Jamison "James" Green, M.F.A. (English/creative writing), is a writer and gender diversity consultant. From March 1991 to August 1999 he was the driving force behind FTM International, Inc., the world's largest and longest-running informational and educational organization for and about female-to-male transgender and transsexual people (formed as the support group called "FTM" in 1986 by Louis G. Sullivan). Mr. Green's writings, film appearances, and public presentations about gender issues have always emphasized the eradication of fear and shame surrounding transgender and transsexual experience, and have been effective in enlightening non-transpeople as to why they ought to care about gender issues. Born in 1948 in

Oakland, California, Mr. Green is the father of a 15-year-old daughter.

Felicity Haynes has written *The Ethical School* (Routledge, 1998) and *The Art of Argument* (Felafel Press, 1987) and published many articles on ethics, critical thinking, art education, and conceptual change. She lectures in philosophy of education and has been involved in university administration particularly in her years as Dean of Education and Head of the Graduate School of Education at The University of Western Australia. She is Co-Director of the International Foundation for Androgynous Studies.

Katherine Johnson is a doctoral student in the Social and Applied Psychology Group at Middlesex University, U.K.

Dave King is a lecturer in the Department of Sociology, Social Policy, and Social Work Studies at the University of Liverpool, author of *The Transvestite and the Transsexual: Public Categories and Private Identities* (Avebury, 1993), and co-author with Richard Ekins of *GenderBlending* (Routledge, 1996). He has published several articles on transvestism and transsexualism.

Delphine McFarlane is a Ph.D. student at The University of Western Australia, where she works as a Director of the Student Guild. She has also worked as an English and health education teacher, freelance journalist, education researcher and education officer. Her previous research has concerned the academic education of pregnant and parenting teenagers and class and gender issues in nineteenth-century literature. Her current focus is on gender ambiguity in early nineteenth-century English women's magazines. Her involvement in the International Foundation for Androgynous Studies has been a significant catalyst for further exploration of hidden genders and the issues confronting those concerned.

Wayne Martino lectures in education at Murdoch University, Western Australia and has published in the field of gender with a particular focus on masculinities and sexualities. He is currently working on two books with Maria Pallotta-Chiarolli including *The Stuff Boys are Made Of!* (to be published by Allen and Unwin in Sydney). He is also working on an edited collection with Bob Meyenn entitled *What About the Boys?* (Buckingham:OUP). Currently he is involved in exploring how issues of masculinity and sexuality impact on the teaching practices of male teachers and teacher candidates in Canada and Australia.

Tarquam McKenna lectures in drama and art therapy at Edith Cowan University, Western Australia. He directs Playback Theatre, and is a psychotherapist with a wide circle of gay friends. He is one of the co-directors of the International Foundation for Androgynous Studies.

Surya Monro was a doctoral student in the Department of Sociological Studies, University of Sheffield, England.

Sam Dylan More works at the Okazaki National Research Institute in Japan. He is cofounder of the transmen group in Berlin, Germany, and writes articles for several FtM publications. His interests range from biomedical aspects of gender-identity development, sociolinguistics with a focus on Eastern/Western culture comparisons, structure analysis of semiconductor compounds, and legal obstacles to transgender parenting to the role of transgender and intersex protagonists in science-fiction/fantasy and contemporary Japanese literature.

Maria Pallotta-Chiarolli is employed in the School of Health Sciences at Deakin University, Australia, and writes and researches in the issues of ethnicity, gender, sexuality, and HIV/AIDS, particularly in relation to education and sexual health. Her publications include Australia's first AIDS biography, *Someone You Know* (Wakefield Press, 1991, new edition 2000), a compilation of culturally and sexually diverse young women's writings and visuals entitled *Girls Talk: Young Women Speak Their Hearts and Minds* (Sydney: Finch Publishing, 1998), and *Tapestry: Five Generations of Italian Women* (Sydney: Random House, 1999). She is co-researching and co-writing books on boys' education and health with Wayne Martino; undertaking research with women having bisexually active male partners with Sara Lubowitz; and undertaking research with Southeast Asian women and Hepatitis C in Australia with Professor Sandy Gifford in a project entitled "Bad Blood, Bad Livers."

Chris Somers has won an international award from the Saudi Arabian government and the King Abdulaziz University for services to universal education concerning Antarctica and the global environment. S/he has flown solo in a powered aircraft, illustrated several books for major international publishing houses and is trained as an educator and professional photographer. Currently completing post-graduate studies, and presenting exhibitions of his/her photographic artworks in the United States, Chris was a cofounder and very active executive member of the International Foundation for Androgynous Studies.

Ashley Tauchert teaches in the School of English, University of Exeter, England.

The International Foundation for Androgynous Studies Inc. (IFAS)

Four contributors to this book have been active on the International Foundation for Androgynous Studies. Based in Western Australia, it was established to redress the too rigid dichotomy between male and female and reexamine accepted criteria for gender status. It is inclusive, encouraging the participation of all transvestites, transsexuals, and intersex persons, gays, lesbians, and bisexuals. It encourages studies of aspects of endocrinology, urology, pediatrics, surgery, psychiatry, clinical psychology, genetics, gender studies, law, politics, history, and education. The foundation supports and encourages research, dissemination of information about androgynous people, and networking and support for them. This book forms part of its ongoing educational enterprise. Inquiries, donations, and expressions of support should be sent to PO Box 1066, Nedlands, Western Australia. 6909; Ph/FAX [618] 9386 7730; *http://www.ecel.uwa.edu.au/gse/staffweb/fhaynes/IFAS_Homepage.html*

Index

45X0, 32, 222
47XXY, 5–7, 31–33, 215
5-Alpha-Reductase Deficency, 222
46XX, 5, 10, 13
46XY, 5, 10, 13

A

Abercombie, D., 170, 178
adrenal hyperplasia (CAH) or adrenogenital syndrome, 1, 7, 21–23, 193, 197, 215
AEGIS, 213
aggressive masculinity, 97
AIS. *See* Androgen Insensitivity Syndrome
AIS Support Group, 213
Alain, 73
alchemy, 191
Alcoff, Linda, 163, 164
Alexina, 145
Alexis, 72, 74, 78, 79, 80
alienation, 7, 19, 36, 73, 78, 83,
Alvarado, Donna, 2, 15
ambiguous genitalia, 1, 4, 8, 197, 200, 215, 222
American Civil Liberties Union, 204
American Educational Gender Information Services, 213
analysis of discourse, 149, 150
androcentrism, 7
androgen 6, 215
androgen insensitivity syndrome (AIS), 1, 7, 30, 188, 193, 213, 215, 219
androgynes, 12, 14, 30, 135, 157–159, 197, 199, 215

androgynous people, 69, 71, 77, 209, 215
  educational needs of, 205, 208
androgyny, 9, 11–12, 14, 15, 29–31, 38, 160
androphilic, 10, 216
Angier, Natalie, 32, 39
anti-discrimination legislation, 2-4, 8, 210, 211
Anzaldua, Gloria, 90, 91, 115
aposthia, 216
Aristotelian logic, 184, 187
Arnold, A.P., 178
Arthur, 55–56
Ashenden, D., 117
Ashley, April, 127, 140
Atkinson, Leslie, 178
authentic self, 76, 86, 161
autogynephilia, 129, 140, 141, 142, 216

B

Bahlberg, 93
Balkan Sworn Virgins, 216
Bannister, P, 149, 154
Barbara, 73
Barbin, Herculine, 194
Barrett, Michelle, 165
Bassett, Pat, 85
Bates, Denise, 197, 199
beard growth, 1, 64, 188
Beaumont Society, 131, 213
Beavis, C., 103, 117
Behm, D.J., 169, 179
being read, 216

Bem, Sandra, 1, 11, 15, 49
Benestead, Esther, *See* Pirelli, Eben
Benhabib, Selye, 50, 196, 198, 199
Benjamin, Harry, 4, 31, 138, 143, 154, 169, 205, 222
Benjamin, John, 15, 39,
berdache, 12, 216
Berger, Maurice, 50
Berliner, Alain, 205
bigendered, 216
Billings, D. B., 150, 154
binaries, 2, 11, 25, 29, 38, 41, 51, 75, 92, 113, 133–137, 159–160, 172, 181–182, 185, 188, 209
  Art/Nature, 66
  body/mind, 15
  day/night, 181–185
  gay/straight, 99
  gender/sex, 2, 10, 15, 20, 56, 65, 109, 116, 186
  homosexual/heterosexual, 75, 114, 209
  human/machine, 24
  male/female, 5, 11, 12, 15, 24, 83, 162, 166, 169, 181, 185, 209
  masculine/feminine, 11, 38, 46, 75, 169, 187
  nature/nurture, 2, 181
  normal/queer, 15
  real/constructed, 2
  science/culture, 4, 15
binary systems, 1, 125, 164, 168, 174, 181–190
biological sex, 62, 216
birth certificates, 8, 34, 36, 37, 56, 126, 220
birth designated sex, 41–46, 48
bisexuals, 2, 12, 15, 29, 110, 178, 188, 204, 216
Blanchard, R., 129, 140
blank subjectivity, 150
Bolin, Ann, 56, 57, 121, 138, 140, 143, 154
Bordo, Susan, 26

Bornstein, Kate, 1, 15, 49, 77, 84, 137, 140, 158, 159, 163, 164, 197, 199, 221, 225
Boswell, John, 26
boundaries, 23, 25, 30, 39, 71, 79, 119, 127, 133, 169, 184–185, 224
Bradley, Susan J., 92–93, 119, 178, 179, 180
Braidotti, Rosa, 27
brain sex, 40, 178, 216
Brenda, 53
Bromley, Simon, 164
Brown, Lee Anderson, 30, 168, 193–201, 223
Buchbinder, David, 29, 39
Bullough, Bonnie and Vern, 49, 141, 199, 201
Burman, E., 149, 154
Burr, Vivian, 149, 154
butch, 48, 69, 85, 173, 178, 216
Butler, Judith, 2, 11, 16, 21, 23, 24, 26, 27, 29, 39, 41, 50, 78, 79, 83, 84, 88, 90, 91, 92, 95, 98, 115, 143, 151, 154, 160, 164, 167, 169, 170, 179 186, 190

C

CAH. *See* adrenal hyperplasia
Califia, Pat, 196, 197, 199
Cameron, Loren, 146, 154, 158, 164, 165
camp, 104, 216, 217
Candice. *See* Terry.
Carla/Rob, 52–53
Cartwright, A., 27
Cartwright, Lisa, 27
castration, 10, 56, 125
Centre for Hormone Research in Australia, 213
Chase, Cheryl, 139
Chauncey, George, 195, 199
Chen, K.H., 85
Chodorow, Nancy, 43, 50

chromosomes, 1, 5, 7, 8, 13, 32, 218, 220, 221
Cicourel, A.V., 170, 179
Cixous, Hélène, 181, 184, 190
clitoral hypertrophy, 10
clitoridectomy, 218
clitoris, 2, 16, 126, 216, 223, 224
clitoromegaly, 215, 216
clitoroplasty, 216
Clément, 190
Cloe, 99
Coates, Susan, 92, 95, 101, 115
Cohen, Ira, 91, 115
Colapinto, John, 3, 16, 37, 39, 70, 199, 200
Colker, Ruth, 110, 115
compulsory heterosexuality, 10, 18, 75, 85, 87, 99, 106, 107, 114, 210, 216
compulsory visibility, 90
Congenital Adrenal Hyperplasia, CAH. *See* adrenal hyperplasia
Connell, Robert, 27, 88, 98, 101–102, 115, 117, 158, 165
Cornell, Drucilla, 50
Coventry, Martha, 206
Cowell, R., 138, 140,
Cranny-Francis, Anne, 11, 16
Crosier, Louis, 84
cross-dressers, 9, 17, 18, 41–50, 51–54, 57, 85, 123, 131, 157, 201, 216, 225, 226
cross-dressing, 60, 109, 112, 116, 130, 132–123, 140, 142, 165, 201, 216
crossing boundaries, 23, 67
cross-living, 216
Crowhurst, Michael, 115
cryptorchidism, 216
Crystal, David, 170, 179
Curran, Greg, 115
Cussen, Jacqui, 87, 91, 110, 113, 115, 116
cyborg, 20, 24, 27, 39, 165
curriculum, 112, 206, 209, 210

D

Dachsler, Simon, 162
Daly, Mary, 30
Damian, 73, 78, 79
Dana-Tabet, Adrianne, 18, 51–58, 223
Dancer, Tamsin, 106, 116
Daston, Lorraine, 189, 190
Dave, 101–104
David, 21–25
Davis, S.N., 84, 154
Denny, D., 139, 140
DES, diethyl stilbestrol, 217
Dessloch, Simon, 1583, 159
de Saussure, Ferdinand, 181, 191
Devor, Holly, 50
Diagnostic and Statistical Manual of Mental Disorders, 169, 176, 178, 217
Diamond, Milton, 3, 7, 9, 16, 70
dichotomies, false 62, 169. *See also* binaries
dick-chicks, 85
differentiation, 10, 184–188
Digby, Tom, 70
di Leonardo, Micaela, 39
disciplinary power, 92
discourse, 3, 6, 8, 143, 148, 193
  legal, 146, 163
  masculine, 153
  medical, 143–146
  religious, 163
  sociocultural, 143–145
discourse analysis, 152, 153
Dixon, C., 88, 102, 116
Docherty, Thomas, 191, 193
Double-X Syndrome, 32
Douglas, Peter, 114, 117
Down's Syndrome, 34
Dowsett, G., 117
Doyle, James A., 60, 70
drag, in, 158
drag queens, 41, 69, 77, 874, 86, 96, 104, 173, 217

drag kings, 69, 142
Dreger, Alice, 2, 4, 5, 12, 16, 32, 39, 206, 210, 225
DSM-IV. *See* Diagnostic and Statistical Manual of Mental Disorders
dualisms. *See* binaries
Due, Linnae, 163, 165
Dupré, John, 190
dykes, 85, 106, 158

E

Ehrhardt, 29
Eiseley, Loren, 121
Ekins, Richard, 9, 29, 39, 40, 121, 123–142, 125, 128, 130, 140, 145, 154, 158, 165, 167, 223,
Elan-Cane, Christie, 133–136, 140, 142, 159, 161, 224
electrologists, 30
electrolysis, 130, 131, 217
Eliade, Mircia, 38, 39
Eliason, M.J, 83, 85
Elisofon, Eliot, 38, 40
embodied gender, 2, 9, 12, 24, 97, 188
endocrine therapy, 10
endocrinologist, 22, 195
en femme, 217
en homme, 217
epicene, 26, 217
Epstein, Debbie, 88, 98, 101, 102, 116, 120, 121, 160, 165, 166, 190
erotic orientation, 10, 35, 53, 60, 67, 95, 110, 128–130, 158, 169, 206, 217, 220, 221
essentialism, 1, 3, 12, 14, 15, 17, 23, 25, 29, 30, 33, 36, 37, 41, 45, 47, 64, 65, 66, 69, 90, 91, 94, 112, 127, 160 167, , 187, 188, 191, 192, 195, 224
estrogen, 13, 217, 219
ethnomethodology, 125
Eugine, Toinette, 38, 39
eunuchs, 133

excitable speech, 21
exclusion, 1–3, 9, 23, 30, 53, 59, 160–161, 205, 209, 210

F

Fa-fa'fine, 217
fags, 85
fairies, 82, 86
Fallowell, D., 127, 140
Fausto-Sterling, Anne, 1, 2, 4, 16, 139, 141, 218, 221
Feinberg, Leslie, 29, 36, 39, 50, 137, 138, 141, 163, 165, 167
female, 16, 21, 88, 94, 110, 111, 217
female penis, 16, 34
female pseudo-hermaphroditism, 217
female-to-male transsexual, 4, 44, 64, 67, 153, 161, 162, 176, 217
femininity, 30–31, 48–49, 53, 60, 62, 64, 65, 66, 89, 92, 94–101, 103, 106, 117, 127, 132, 136, 138, 147, 152, 158, 186, 187, 189, 194, 217, 218, 220
feminist theory, 7, 16, 20, 23, 24, 29, 40, 41, 43, 48, 59, 153, 157, 159–166, 187
femmes, 69, 217
ferm, 217
fetishistic transvestite, 18, 218
Fotos, S., 177, 179
Foucault, Michel, 4, 6, 16, 32, 39, 51, 73, 84, 87, 88, 90, 108, 116, 143, 148, 150, 154, 183, 194, 199
Fradenburg, Louise, 189, 190
Frank, B., 98, 99, 103, 116
freaks, 14, 34, 38, 110, 160, 167, 196, 197
Freccero, Carla, 189, 190
FtM Transgender Guide, 213
fuzzy logic, 12–14, 194
fuzzy logic gender, 185, 186

G

Garber, Marjorie, 29, 39, 41, 50, 111, 116, 160, 165
Garfinkel, H., 124, 137, 141
Garry, A., 39
Gasche, Rodolphe, 39
Gatens, Moira, 27
gay, 17, 29, 35, 55, 71–79, 81, 82–86, 98–109, 115, 117, 120, 121, 128, 132, 149, 158–159, 177, 203–205, 207–211, 208, 210, 214, 218
gay bashing, 85, 175, 203–204
Gay Lesbian and Straight Education Network, 213
gays, 2, 4, 9, 10, 12, 29, 149, 222
gay youth, 207
gaze, the, 7, 10, 32, 83
Gearhart, John 2
gender, 1–5, 8–9, 19, 29, 32, 35, 41–49, 55, 59–70, 87–120, 143, 158, 171, 218, 226
gender ambiguity, 19
gender and language, 66
gender as native language, 173
gender atypicality, 31, 113
gender benders, 86, 94
gender blending, 121
gender blindness, 183
Gender Centre (Australia), 213
gender community, 218
gender conformity, 26
gender duality, 10–13, 89, 196
gender dysphoria, 4, 10, 18, 36, 170, 194, 218. *See also* gender identity disorder
gender fluidity, 14, 20, 168, 195, 199
gender freedom, 122,
genderfuck, 218
gender identity disorder, 2, 37, 92, 95, 98, 101, 107, 112, 114, 117, 119, 138, 141, 182, 218
gender multi-culturalism, 116

gender neutral 218
gender outlaw, 1, 15, 45, 137, 218
gender performativity 151
gender policy for schools, 207–209
gender reassignment, 112, 218, 221
gender role, 20, 218, 221
gender, the third, 11, 14, 38
Gendys, 213
Gender and Sexuality Alliance, 159
genes, 13, 32
Genetic Alliance, 213
genetic sex, 218, 221
genetic testing in sport, 8
genital "correction", 111, 200
genital mutilation, 7, 197
genitalia, 2–4, 8–10, 13, 14, 19 , 24, 32, 37, 67, 68, 144–145, 198, 200, 216, 217, 225
genitoplasty, 218
Gergen, K., 151, 154
Gergen, M.M., 84
GID, *See* gender identity disorder
Giddens, 87
Gilbert, G.N., 118, 148, 154
Gilbert, M.A., 3, 18, 41–50, 223
Gilbert, S.M. 190
GIRES, 213
Goffman, Erving, 47, 50, 77, 84
Goldschmidt, J., 195
gonadal dysgenesis, 1, 10, 218
Gonzalez, Jennifer, 20
Goodman, J., 83, 84
Goss, R., 84
Gray, Chris Hables, 27
Green, Jamison, 17, 18, 59–71, 94, 223
Green, Richard, 118
Greenberg, Julie A., 7, 16
Grosz, Elizabeth, 148, 155
Grote, J., 195
gynocophilic, 10, 218
gynogenesis, 24, 27, 218

H

haciendo caras, 91–92. 218
hair, 6, 8, 13, 126, 128–132, 134, 188, 215
Halberstam, Judith, 132, 137, 141, 142
Hale, C. Jacob, 65, 70
handholding, 79–80
Hall, E.T., 175, 177, 182
harassment, 92, 97, 99–102, 108–110, 203–204, 207–208
Haraway, Donna, 20, 24, 25, 27, 29, 39, 160, 165
Harrison, David, 158
Harry Benjamin International Gender Dysphoria Foundation, 138, 154, 169, 222
Harter, S., Waters, P.L., and Whitesell, N.R., 83, 84
Hawking, Stephen, 34
Haynes, Felicity, 1–16, 10, 17, 29–40, 115, 224
Haywood, C., 102, 103, 116
Hazel, 54–55
Healy, A, 117
Heilbrun, Carol, 38, 39
Heilman, E., 83, 84
Hekma, Gert, 199, 200
Helena, 100, 131–132
Henriques, 144, 151, 153, 155
Herdt, Gilbert, 16, 60, 70, 140, 199
herm, 218
hermaphrodite, 4, 9, 10, 14–16, 36, 109, 110, 111, 189, 194–196, 198
Hermaphrodite Education and Listening Post, 213
Hermaphrodites with Attitude, 5
hermaphroditism, 4, 12, 219
Heron, Leonard, 34, 40
heteronormative practices, 74, 87, 203
heteronormative regimes, 88, 169, 209
heteronormativity, 22, 92, 94, 98, 102–104, 110, 114, 167, 209–211, 219
heteropatriarchy, 122, 164, 200
heterosexism, 93, 207–209
heterosexuality, 2, 22, 23, 92, 93, 100, 102–106, 109–114, 117, 132, 134, 148, 149, 163, 201, 209, 223
Himmelwait, S., 161
Hippocratic tradition, 192
Hirschfield, 195
Hirst, P., 89, 116
Hite, S., 99, 103, 116
*hjira*s, 12, 117, 118, 219
Hmong immigrant, the, 9–10
Hoenig, J., 144, 160
Holland, J.C., 103, 116
Hollway, Wendy, 150, 151, 155
Holmes, Morgan, 36, 116
homophobia, 72, 78, 98, 100, 105, 109, 117, 118, 203–204, 207–210, 213, 219
homophobia, internalized, 71, 72, 219
homosexual, 2, 10, 12, 15, 60, 95, 103, 104, 112, 116, 117, 120, 121, 141, 169, 177, 197, 198, 203, 220, 223
homosexuality, 173, 180, 219
homovestite, 219
hooks, bell, 159, 165
Horowitz, I.L., 141
Hooley, Jillian, 112, 116
hormonal treatment, 3, 4, 6, 9–10, 37, 52, 219, 223
hormones, 29. 32, 54, 129, 138, 219
Hunter, I., 89, 116
hypogonadotrophic hypogonadism, 188, 222
hypospadias, 1, 197, 215, 219
hypothalamus, 13, 32
hysterectomy, 10, 22, 56, 125

I

identity, 4, 8, 9, 14, 15, 20, 24, 26, 36, 44–46, 51, 62, 74, 77, 83, 88,

Index 233

90–93, 112–115. 121, 124, 128, 138, 195, 205, 209, 219
body, 158
gender, 4, 9, 10, 13, 19, 20, 24, 51, 54, 62, 67, 68, 94, 95, 135, 140, 157–165, 167, 169–178, 191, 218
transgender 13, 51–57, 121, 138
transsexual, 149
incarnation, 2, 80
inclusivity, 207
incorrigible propositions, 62, 70
indexing in communication, 170–176, 178
International Bill of Gender Rights, 138, 204
International Federation of Klinefelter's Syndrome Support Groups, 213
International Foundation for Androgynous Studies, 9, 213, 225,
International Foundation for Gender Education, 52, 54, 213
International Journal of Transgenderism, 213
interpretative repertoires, 149
intersex, 1, 2, 4, 5–9, 12, 15, 16, 29–38, 67, 114, 139, 157, 158, 160–162, 163, 169, 173, 174, 176, 178, 193–201, 204, 205, 207–21185, 190, 191, 193, 195, 196, 199, 201, 207, 208, 210, 219
Intersex Genital Mutilation, 7
Intersex Society of Australia, The, 213
Intersex Society of North America, The, 7, 139, 213
Intersex Support Group International, 213
intersexuality, 87, 109, 139, 163, 193–201, 219,
invisibility, 11, 26, 73, 76, 78, 81–83, 87, 90, 196, 205, 209
ISGI, 213
It's Time Illinois, 213
Iversen, Mark, 207

J

Janice, 129–130
Jason, 104–106
Jay, 71–72, 74, 75, 76
Jessel, D., 32, 40
Joanna/Dave, 158, 162
John, 132–133
John/Joan, 3, 7, 37, 70, 200
Johns Hopkins University, 2, 40,
Johnson, Katherine, 121, 144–155, 224
Jordan, 99, 117
Jorgenson, Christine, 195–196, 197, 201

K

Kacprzyk, Janusz, 191
Kallmann's syndrome, 219
Kando, 145, 155
karyotypes, 8, 224, 220
Kate, 95
kathoey, 111, 220
Kazmi, Yedullah, 90, 117
Kehily, M., 88, 98, 102, 103, 117, 118
Kenway, Jane, 1, 16
Kessler, Suzanne, 5, 16, 62, 63–64, 67, 68, 70, 117, 124–126, 141, 143, 155, 219
Kimmi, 53
King, Billie Jean, 8
King, Dave, 4, 9, 16, 29, 39, 40, 123–142, 143, 154, 165, 171, 224
Kinsey scale, 220
kissing, 80
Kitzinger, C., 143, 155
Klinefelter, H.F., 39
Klinefelter Syndrome, 5–6, 8, 11, 12, 31, 39, 219, 220
Klinefelter's Syndrome Association, 213
Knobel, M., 117
Kosko, Bart, 190, 186

Kulick, Don, 59, 70
Kumar, Arvind, 111, 117

L

labeling, 2, 14, 32, 59, 61–62, 92, 97, 115, 160, 169, 195, 206, 220
Lacan, J., 77, 84
Laclau, Ernesto, 190, 191
Lancaster, Roger, 39
Langley, Jess, 95, 117
Laqueur, Thomas, 16, 143, 155, 194, 199
Laskey, L., 103, 117
Laver, J., 170, 172, 179
Lawrence, Ann, 129, 141
Lee, Glenda, 36, 40
lesbian identity, 2, 9, 10, 12, 29, 40, 48, 60, 61, 67, 73, 99, 123, 141, 158, 159, 165, 174–178, 203, 204, 205, 207–209, 220
LeShan, Lawrence, 39
Letts, Will, 209, 210
Lev, Arlene Istar and Sundance, 175, 179
Lind, Earl, 195, 199
Lingard, Bob, 114, 117
linguistic codes, 172
Lloyd, Genevieve, 27
Lorber, Judith, 29, 39, 158, 159, 160, 165
Luhrmann, Baz, 79
Lumby, Catharine, 27
Lykke, Nina, 27

M

Mac an Ghaill, Mairtin, 80, 98, 117
Maccoby, E.E., 170, 172, 179
MacKenzie, Gordene, 50
Mahu, 220
Mak nyah, 220
male femaling, 121, 133, 140
male impersonator, 220
male pseudo-hermaphrodite, 220
male-to-female transsexual, 4, 30, 44, 68, 140, 147, 153, 217, 220
Mardi Gras, 29
marriage, 11, 126, 161, 164, 169, 191
Marshall, Susan, 1
Marteene, Janet, 37, 40
Martin, B., 177, 179
Martino, Wayne, 18, 87–118, 89, 95, 98, 99, 114, 117, 224
masculinity, 24, 26, 27, 30, 48, 50, 53, 56, 60–67, 68, 89–115, 126, 176, 185, 186, 220
Mason and Tomsen, 103, 117
mastectomy, 31, 56
Mauss, 88–89, 90, 91, 92, 95, 97, 111, 117
Mayne, Xavier, 195, 200
mAy-welby, norrie, 117, 145, 149, 155, 214
McCloskey, D., 127, 130, 141
McFarlane, Delphine, 17, 19–27, 85, 224
McKenna, Tarquam, 17, 71–86, 128, 203–211, 224
McKenna, Wendy, 62, 63–64, 67, 68, 70, 124–126, 141, 143
McLaren, Peter, 77, 81, 82, 84, 85
McMullen, M., 169, 179
McNay, Louis, 183, 190
medical model, 4, 20, 23, 24, 27, 31–37, 51, 56, 63, 110, 112, 115, 127, 138, 143–145, 169, 194
metoidioplasty, 220
Michael, 43–44
micropenis, 2, 215, 220
migrating, 121, 125, 127–130, 131, 137, 138
Miller, Bethwyn, 95, 117
Mills, C. Wright, 141
mince, 85
Minter, Sharon, 93, 101, 113, 117
Moir, A., 32, 40
Money, John, 2, 3, 29
Monro, Surya, 121, 157–165, 168, 224

Index

More, Kate, 158
More, Sam Dylan, 18, 41, 68, 99, 168, 169–180, 182, 224
Morley, D., 85
Morris, Jan, 113, 141
Morris, Rosalind, 111, 118
Mulkay, M., 148, 154
multiple subjectivity, 151
Murdoch Children's Research Institute, 213
Murphy-Shigematsu, S., 177, 179
Murumcrew, C., 179
Muxe, 220

N

Nabozny, Jamie, 203
nancies, 85
Nanda, Serena, 111, 118
Nataf, Zach, 138, 141, 158, 160, 163, 165
National Society for Genetic Counsellors, 213
Nayak, A., 88, 98, 102, 103, 118
negating, 125, 133–136
Neisen, 93, 101, 115, 118
neuter, 220
neutral pronouns, 135–136, 167
New Zealand Sex Chromosome Society, 213
New Zealand Transsexual Men's and Women's Resources, 213
Nicholson, Linda J., 26
Noonan's Syndrome, 220
No-op, 220, 221
norms, 2, 4, 5, 7, 9–11, 23, 26, 32, 34, 37, 53, 59, 103, 114, 160, 114, 203, 205, 207, 226

O

objectivity, 25
Olympic Games, 8
order of discourse
  modes of desire, 82

modes of production, 82
modes of subjectivity, 82
Orlandic principles, 114
Orlando, 19, 20, 26, 113
Ornstein, Robert, 38, 39
Osborne, Harold, 17
Osburg, S., 169, 170, 179
oscillating, 123, 127, 132, 133, 134, 135, 140
osteoporosis, 31, 34
Oudshoorn, Nelly, 27
Outreach Institute of Gender Studies, 214
ovaries, 13
over-compensation, 175

P

Pallotta-Chiarolli, Maria, 18, 87–118, 94, 109, 111, 114, 117, 118, 225
Paludi, Michele A., 60
panopticonic surveillance, 92, 94, 110
pansexuality, 158
Park, Katherine, 189, 190
Parker, A., 102, 103, 118, 149, 150, 154, 155
passing, 68, 91, 176, 178, 221
pathologizing, 1–4, 9, 11, 32, 54, 94, 100, 162, 170, 210
patriarchy, 18, 30, 36, 70, 163, 170, 173
Pauline, 131, 132
Pearsall, M., 39
Pendrey, Catherine, 95, 118
penis, 3, 5, 6, 9, 16, 13, 21, 23, 32, 34, 53, 56, 64, 67, 126, 131–133, 135, 176, 200
Penley, Constance, 27
Pepper, J., 131, 141
per, 221
performance, 92, 137, 169
  linguistic, 172, 176–177
performativity, 13, 71, 90, 91, 93, 96, 100, 103, 156, 160, 168, 170, 221

P.E.R.S.O.N. Project, The, 214
phallocentric hegemony, 20
phalloplasty, 56, 153
phallus, 161
phenotypes, 31, 215
physical sex, 221
Pirelli, Eben/Benestead, Esther, 205, 210
Plato, 195
Plummer, Kenneth, 91, 118, 121, 124, 127, 137, 141, 144, 155, 158, 160, 165, 200
poofter, 79, 86, 94, 104
positioning, 151, 188
Post-op, 221
Potter, Jonathan, 148, 151, 152, 155
Pratt, Minne Bruce, 111, 118
pregnancy, 107
    termination of 35–36
Pre-op, 221
Press for Change, 146, 214
Price, Deb, 204, 210
Pride, Gay, 159
Pride London, 214
Pridenet, 214
Pride NY Gay Lesbian Web Directory, 214
primary sex characteristics, 221
Prince, Virginia, 52, 54, 57, 130, 141
prison, 162
progesterone, 221
progestin-induced virilization, 221
pseudo-hermaphroditism, 31, 200,
psychiatrists, 2, 9, 30, 162
psychologists, 25, 30, 67, 195
psychopathology, 94
puberty, 5, 21, 127, 128, 136, 215, 219
public toilets, 11, 164
Pursey, Belinda, 100, 118

Q

queer, 2, 3, 4, 9, 10, 12, 15, 29, 37, 71–74, 79, 85, 106, 107, 112, 116, 124, 160, 169, 175, 179, 182, 188, 221
queer sexualities, 188, 221
queer theory, 11, 23, 160
Query, Julia, 20, 27

R

Rabinow, Paul, 90, 118
Rajchman, J., 88, 118
Ramazangolu, C., 118
Ramet, S., 124, 141
Ratti, R., 119
Raymond, Janice, 9, 16, 29–30, 35, 40, 162, 165
read, 6, 7, 172, 174, 221
real life test, 221
Redman, D., 98, 101, 102, 118
Reimer, David, 3
Renaissance Transgender Association (Inc.), 214
rent boys, 80
repression, 152
reproductive organs, 4, 10, 13, 14, 32, 221
Rhian, 35
Rich, Adrienne, 75, 85
Richards, Renee, 8
Riddell, Carol, 30, 40
Roberto, L.G., 144, 145, 155
Roberts, Gareth, 191
Rogers, Michelle, 99, 100, 118
Romeo, 78–79
Rose, Debra, 134–135, 141–142
Rottnek, Matthew, 93, 98, 117, 118
Rowe, R., 131, 142
Rubin, Gayle, 163
Rubin, Henry, 153, 155

S

Sally, 100–101, 118
Sapphists, 193
Schwartz, 40

Science, 24–25
Scott, 98–99, 106, 107
Scott, Joan W., 11, 16
Scott, Lyn, 115
script evasion, 91
script switching, 91
Seabrook, Laura Ann, 109, 119
secondary sex characteristics, 7, 8, 221
Seahorse, The, 52
Segal, L., 151, 152, 155
self, 19, 32, 77, 170, 172, 176, 196, 199
self-ascription, 87, 88, 90, 94, 98, 147, 161, 196
Sewell, 127–128, 142
sexology, 194
sex reassignment surgery, 16, 39, 52, 220, 221
sexuality, 8, 17, 21, 24, 25, 26, 29, 32, 38, 50, 69, 72, 87, 91–92, 93, 94, 98, 99, 100, 101, 110–12, 115–120, 123, 125, 130, 131, 134–137, 142, 143, 155, 161, 163, 166, 168, 193, 207, 210, 214, 223
sexual preference *or* sexual orientation, 2, 10, 13, 71, 169, 177, 178, 214, 221
Shapiro, Judith, 183, 193
Sharpe, S., 118
she-males, 16, 56, 86, 165, 222
Sigmundson, H. Keith, 3, 16
Signorile, M., 104, 119
Singer, June, 38, 40
sissies, 34, 45, 86, 99, 134–135, 142, 173
sissy boy, 222
sissy maid, 134
Sissy Maid Quarterly, 134, 142
Skelton, C., 102, 119
Slack, J., 75, 85
Smith, Heather, 94, 119
Snitow, Ann, 75, 85
Social Construction Theory, 18, 60
socialization, 41, 42, 43, 44, 45, 46, 47, 48, 49, 52
sodomites, 85
Somers, Chris, 7, 10, 17, 29–40, 85, 225
Somers Syndrome, 32
Sourbut, Elizabeth, 20, 24, 27
splitting, 152
Spry, J., 126, 142
standards of care, 222
Stansell, Christen, 85
Steinberg, 101, 103, 119
Stella, 147
Stephen, 91, 95–98, 108–110, 111, 112, 115
stereotypes, 8, 15, 30, 38, 64, 106, 107, 116, 162, 164, 165, 211
stigma, 206
Stoller, Robert, 143, 150, 155
Stone, Sandy, 20, 24, 26, 27, 49, 70, 160, 1651, 194
stories, 2425, 26, 32, 116, 191, 197
  intersex, 141
  migrating, 139
  oscillating, 140
  transcending, 139, 141, 142
straight-acting, 105
Straub, K., 160, 165, 171, 193
Stryker, Susan, 160, 165
subjectification, 89, 183
subjectivity, 144, 182–183, 186, 188, 197
Sudnow, D., 181
Sullivan, Louis G., 223
surgery, 1, 3, 7, 9, 10, 15, 16, 21, 25, 30, 34–50, 51, 55, 64, 110, 126, 128, 130, 136, 139, 161, 175, 176, 188, 193, 198, 200, 203, 206
Susan, 80–81
symbolic marking, 145–146, 200
symptoms, 144

T

Tardieu, Auguste, 194

Tasker, Yvonne, 27
Tauchert, Ashley, 11–13, 168, 181–191, 225
Taylor, Laurie, 91, 117
Taylor, M., 154
technologies, 20, 24, 27, 65, 124, 127, 153, 164, 170, 190
Terry, 76, 77, 78
testes, 6, 13, 219, 223
testicular feminization, 215, 222
testosterone, 1, 6, 7, 13, 31, 37, 215
textual violence, 20–22, 26
third gender, 38, 54, 136, 138, 197, 199
Third International Congress on Sex and Gender, 1, 7, 40, 70, 85
third sex, 140, 158–159, 195, 200
Thomas, S., 142
Thompson, R., 127–128, 142
Thompson, Sharon, 85
Tindall, C., 154
tomboy, 45, 95–97, 117, 173
tomgirl, 87, 95, 98, 109
Torrai, M., 173, 179
tranny, 222
transcending, 125–126, 136–139, 140
transgendering processes, 123–142
   concealing, 126
   erasing, 125
   implying, 126
   redefining 126
   substituting, 125
transgenderist, 130
transgender persons, 18, 51–56, 63, 87, 113, 157–164, 167, 188, 202, 207, 222
Transgender Australia, 214
Transgender London, 159
Transgender Tapestry, 56, 223
trangressive gender, 98
transition, 226
transitioning, 52
transphobia, 122, 162–163, 222
transsexuality, 36, 40, 113, 188, 216, 219
Transsexual Menace, 145, 146, 170, 214
transsexuals, 3, 4, 9, 10, 12, 17, 24–26, 29, 30, 41, 46, 51–59, 61, 65, 67, 113, 124, 129–142, 143–150, 154, 157, 166, 170, 222
transsexualism, 143, 214
transvestism, 219
transvestites, 2, 4, 12, 29, 41–50, 52, 53–56, 86, 111, 123, 126, 131, 133, 134, 135, 140, 143, 144, 151, 157, 162, 219, 220, 222
Treichler, Paula, 27
Trumbach, Randolph, 189, 191
Turner Syndrome, 21, 188, 193, 219, 222
Transsexual Women's Resources, 214

U

Ueda, Y., 177, 179
Ulrichs, Karl Heinrich, 195
Ultrasex, 214
UK Intersex Association, 214
Understanding Klinefelter Syndrome, 214
Universal Declaration of Human Rights, 203, 210
unseen genders, 15, 17, 19, 24
Uranian, 195
Urban, 150, 154
uterus, 1, 6, 21–22, 32

V

vagina, 6, 9, 13, 21, 23, 30, 31, 32, 37, 45, 67, 126, 129, 133
vaginal reconstruction, 9
vaginoplasty, 52, 132
Vance, Carole, 50, 152, 155
Volcano, Danny, 132, 142
von Mahlsdorf, Charlotte, 123, 142

W

Wagner, Willie, 204
Walby, Sylvia, 165
Walker, J., 101, 103, 119
Walkerdine, Valerie, 155
Walters, William, 193, 200
Ward, N., 98, 119
Waters, P.L., 85
Watts, Alan, 38, 40
Webb, John, 27
Webb, T./R., 142, 128–129
Wedge, B., 179
Weedon, C., 146, 155
Weil, Kari, 29, 40
Weinstein, E.A., 182
Weize, C., 169, 170, 179
Wetherell, Margaret, 148, 151
Whinnom, Alex, 163
Whitesell, N.R., 83, 84
Whittle, Stephen, 160, 161, 162, 165, 169, 178, 179, 180
Wickelgren, W.A., 170, 180
Wilchins, Riki, 137, 142, 172, 180, 197, 200
Wilde, Oscar, 83
Williams, Saree, 108, 119
Willis, P., 101, 103, 119
Willis, Sue, 1, 16
Wintke, 222
Winyanktecha, 222
Wittgenstein, Ludvig, 12, 16, 168
Wittig, Monique, 50
woman, 1, 6,8, 31, 35, 36, 37, 52, 59, 61, 64, 67, 87, 102, 111, 118, 123, 142, 158, 182, 186, 188, 196, 199, 222
Wood, J., 103, 119
Woodward, Kathryn, 40, 145, 147, 152, 155
Woolf, Virginia, 20, 29, 113, 119
Woolley, P., 89, 116

Y

Youth Pride Alliance, 214

Z

Zadeh, Lofti, 191
Zucker, Kenneth J., 92–93, 95, 119, 178, 180

 **ERUPTIONS**
New Thinking across the Disciplines

Erica McWilliam
General Editor

This is a series of red-hot women's writing after the "isms." It focuses on new cultural assemblages that are emerging from the deformation, breakout, ebullience, and discomfort of postmodern feminism. The series brings together a post-foundational generation of women's writing that, while still respectful of the idea of situated knowledge, does not rely on neat disciplinary distinctions and stable political coalitions. This writing transcends some of the more awkward textual performances of a first generation of "feminism-meets-postmodernism" scholarship. It has come to terms with its own body of knowledge as shifty, inflammatory, and ungovernable.

The aim of the series is to make this cutting edge thinking more readily available to undergraduate and postgraduate students, researchers and new academics, and professional bodies and practitioners. Thus, we seek contributions from writers whose unruly scholastic projects are expressed in texts that are accessible and seductive to a wider academic readership.

Proposals and/or manuscripts are invited from the domains of: "post" humanities, human movement studies, sexualities, media studies, literary criticism, information technologies, history of ideas, performing arts, gay and lesbian studies, cultural studies, post-colonial studies, pedagogics, social psychology, and the philosophy of science. We are particularly interested in publishing research and scholarship with international appeal from Australia, New Zealand, and the United Kingdom.

For further information about the series and for the submission of manuscripts, please contact:

> Erica McWilliam
> Faculty of Education
> Queensland University of Technology
> Victoria Park Rd., Kelvin Grove Q 4059
> Australia

To order other books in this series, please contact our Customer Service Department at:

> (800) 770-LANG (within the U.S.)
> (212) 647-7706 (outside the U.S.)
> (212) 647-7707 FAX

Or browse online by series at:
> www.peterlang.com